# Tribology
## Principles and Design Applications

R. D. Arnell
P. B. Davies
J. Halling
T. L. Whomes

# M
## MACMILLAN

First edition 1991

Published by
MACMILLAN EDUCATION LTD
Houndmills, Basingstoke, Hampshire RG21 2XS
and London
Companies and representatives
throughout the world

Typeset in the UK by P&R Typesetters Ltd,
Salisbury, Wilts.
and Printed in Hong Kong

British Library Cataloguing in Publication Data
Tribology.
1. Tribology
I. Arnel, Derek
621.89
ISBN 0-333-45867-2
ISBN 0-333-45868-0 pbk

# Contents

# Preface

In 1975 we and four of our former colleagues wrote the textbook *Principles of Tribology*. That book was very successful both in the United Kingdom and overseas and it is still being sold today. However, 13 years after the first publication of the book, the publishers suggested that we should consider revising the text, to include the more important developments which had taken place in the meantime.

Although we have remained active in tribological research, it was only when we started work on this revision that we realised that the developments which have taken place have been so extensive that they could not be accommodated simply by revision of our existing text. As a result, although we have reused some of our original material, the present book has resulted from a complete reappraisal of our ideas on the necessary content of a modern introductory textbook on tribology. As the title of the book implies, it has been our intention to write a book which will provide a thorough grounding in the principles of our subject for undergraduates and new research workers and will also act as a basic reference book for practising designers.

In the areas of friction and wear, although there has been a great deal of high-quality research on the underlying physical and mechanical processes, there are no current, or indeed foreseeable, models which will allow quantitative predictions of friction coefficients or wear rates from basic physical, chemical and mechanical properties. Such predictions will continue to be based on experimental measurements. Therefore, in these areas, we have attempted to set out as clearly as possible the underlying processes and thus to assist the reader in making sensible initial choices of materials.

In the areas of fluid film lubrication, as well as covering the fundamental mechanisms, we present simple design procedures for hydrodynamic bearings and incorporate the more important aspects of the growing amount of information on elastohydrodynamic lubrication, with examples of its use in design.

We have also included in the book introductions to more practical considerations such as lubricant selection, lubricant-supply systems, condition monitoring and failure analysis.

We should like to record our thanks to Mrs Marise Ellidge for her help in preparing the typescript and to Miss Jenny Shaw, formerly Engineering Editor of Macmillan Education, for her great tact and patience.

*May, 1990*

R. D. A.
P. B. D.
J. H.
T. L. W.

# Notation

| Symbol | Definition |
|--------|------------|
| $A$ | Real area of contact |
| $A_F$ | Area of filter |
| $a$ | Semiaxis of contact ellipse in transverse direction, or radius of contact circle (Chapters 1, 7) |
| $a$ | Width of axial supply groove (Chapter 6) |
| $B$ | Width of bearing |
| $b$ | Length of journal bearing (Chapter 6) |
| $b$ | Semiaxis of contact ellipse in direction of motion (Chapter 7) |
| $C$ | Dynamic load capacity of rolling bearing (Chapter 8) |
| $C_0$ | Static capacity of rolling bearing |
| $C$ | $k_r/k$ |
| $C_p$ | Specific heat |
| $C_d$ | Diametral clearance |
| $c$ | Radial clearance |
| $D$ | Diameter of gear wheel |
| $d$ | Diameter of journal bearing |
| $d_m$ | Mean diameter of rolling bearing |
| $E$ | Elastic modulus |
| $E'$ | Elastic modulus of contact $= 2 \left/ \left( \dfrac{1-v_1^2}{E_1} + \dfrac{1-v_2^2}{E_2} \right) \right.$ |
| $E^*$ | Elastic contact modulus $= 1 \left/ \left( \dfrac{1-v_1^2}{E_1} + \dfrac{1-v_2^2}{E_2} \right) \right.$ |
| $e$ | Shaft eccentricity |
| $F$ | Shear or friction force |
| $F_a$ | Axial force |
| $F_R$ | Rolling friction |

| | |
|---|---|
| $F_{rad}$ | Radial force |
| $F_S$ | Sliding friction |
| $F_x, F_y, F_z$ | Force in coordinate directions |
| $\bar{F}$ | $F/(\eta U_0 BL/h_1)$ |
| $f$ | Fraction of scratched volume to become wear debris |
| $\hat{G}$ | Materials parameter $(= E'\alpha)$ |
| $g$ | Acceleration due to gravity |

$$g_e \quad = \frac{\hat{W}}{\hat{U}^{1/2}} \qquad = \frac{\hat{W}^{8/3}}{\hat{U}^2}$$

$$g_h \quad = \frac{\hat{W}}{\hat{U}}\hat{H} \quad \left.\begin{array}{c}\text{rectangular}\\\text{contact}\end{array}\right\} \quad = \left(\frac{\hat{W}}{\hat{U}}\right)^2 \hat{H} \quad \left.\begin{array}{c}\text{elliptical}\\\text{contact}\end{array}\right\}$$

$$g_v \quad = \frac{\hat{W}^{3/2}}{\hat{U}^{1/2}}\hat{G} \qquad = \frac{\hat{G}\hat{W}^3}{\hat{U}^2}$$

| | |
|---|---|
| $H$ | Hardness (Chapters 2, 3) |
| $H$ | Power to rotate shaft (Chapter 6) |
| $\bar{H}$ | Dimensionless power coefficient |
| $\hat{H}$ | Dimensionless film thickness $(= h/R)$ |
| $h$ | Dimensionless surface separation (Chapter 1) |
| $h$ | Film thickness (Chapters 6, 7, 8) |
| $h_c$ | Central film thickness |
| $h_m$ | Film thickness at maximum pressure |
| $h_{max}$ | Maximum film thickness in a journal bearing |
| $h_{min}$ | Minimum film thickness in a journal bearing |
| $h_0$ | Minimum film thickness |
| $h_p$ | Film thickness at pitch line |
| $h_s$ | Starved film thickness |
| $h_1$ | Outlet film thickness of slider |
| $h_2$ | Inlet film thickness of slider |
| $h_\infty$ | Film thickness with fully flooded inlet |
| $I$ | Inlet shear heating parameter $= \left(\dfrac{\eta_0 \delta U^2}{K}\right)$ |
| $J$ | Inlet factor $(= bx_i/(2Rh_0)^{2/3})$ |
| $j$ | Coordinate in $X$ direction |
| $K$ | Wear coefficient (Chapter 3) |
| $K$ | Bearing inclination (Chapter 5) |
| $K$ | Thermal capacity constant (Chapter 6) |
| $K$ | Thermal conductivity (Chapter 7) |
| $K_F$ | Filter characteristic |
| $K_G$ | Grease lubrication constant |
| $k$ | Yield shear stress (Chapters 1, 2, 3) |
| $k$ | Specific wear rate (Chapter 4) |
| $k$ | Proportion of heat convected by oil (Chapter 6) |

| | |
|---|---|
| $k_e$ | Ellipticity parameter ($= a/b$) |
| $k_0$ | Constant specific wear rate |
| $k_r$ | Contaminant shear stress |
| $L$ | Length of sliding bearing (Chapter 5) |
| $L$ | Life of bearing in revolutions (Chapters 3, 8) |
| $l$ | Length of axial supply groove (Chapter 6) |
| $l$ | Centre distance of gears (Chapter 7) |
| $M$ | Mass supported by journal bearing (Chapter 6) |
| $M$ | Moment of frictional force (Chapter 8) |
| $m$ | Dimensionless distance to inlet boundary ($= x/b$) |
| $m_i$ | Critical inlet boundary distance |
| $N$ | Speed (rev/s) |
| $N_F$ | Number of cycles to failure |
| $N_l$ | Number of ordinates of spacing $l$ in sample length |
| $n$ | Exponent in stress–strain equation (Chapter 1) |
| $n$ | Number of contacts (Chapter 3) |
| $n$ | Speed (rev/min) (Chapter 9) |
| $P$ | Load per unit length (Chapter 1) |
| $P$ | Power, rate of heat generation (Chapters 5, 8) |
| $P$ | Dynamic equivalent load on rolling bearing (Chapter 8) |
| $P_0$ | Static load on rolling bearing |
| $p$ | Pressure |
| $p_0$ | Yield pressure |
| $p_f$ | Pressure of fluid in supply groove |
| $p_h$ | Maximum Hertzian pressure |
| $p_{mean}$ | Mean contact pressure |
| $p_{max}$ | Maximum pressure |
| $\bar{p}$ | Normalised pressure ($= p/H$) |
| $Q$ | Tangential force in rolling contact (Chapter 2) |
| $Q$ | Wear volume (Chapter 3) |
| $Q$ | Volume rate of flow (Chapters 5, 6, 9) |
| $\bar{Q}$ | $Q/(Bh_1 U_0)$ |
| $\bar{Q}$ | Dimensionless flow rate |
| $Q_p$ | Pressure-induced flow |
| $Q_v$ | Velocity-induced flow |
| $Q_d$ | Depth wear rate |
| $q$ | Tangential traction in rolling contact |
| $R$ | Equivalent radius of curvature (Chapter 7) |
| $R$ | Relubrication factor (Chapter 9) |
| $R_A$ | Mean roughness of a surface |
| $R_e$ | Reynolds number |
| $R(l)$ | Autocorrelation function |
| $R_x$ | Radius of curvature in direction of rolling |
| $R_y$ | Radius of curvature in transverse direction |

| | |
|---|---|
| $\bar{R}$ | Ratio between plastic and elastic area |
| $R_1$ | Radius of journal bearing |
| $R_2$ | Radius of shaft |
| $r(l)$ | Normalised autocorrelation function |
| $s$ | Dimensionless asperity height (Chapter 1) |
| $s$ | Friction force/unit area (Chapter 2) |
| $s$ | Slid distance (Chapter 3) |
| $T$ | Temperature |
| $T_e$ | Effective temperature of oil |
| $T_f$ | Temperature of oil at feed groove |
| $T_1$ | First transition load |
| $T_2$ | Second transition load |
| $\bar{T}$ | Dimensionless temperature coefficient |
| $t$ | Thickness of surface film |
| $U$ | Rolling speed $(=(U_1 + U_2)/2)$ |
| $\hat{U}$ | Dimensionless speed parameter $(=\eta U/E'R)$ |
| $U_0$ | Velocity of moving surface or shaft surface |
| $U_c$ | Velocity of surface |
| $u$ | Speed |
| $V$ | Scratch volume |
| $V_p$ | Velocity at pitch line |
| $V_e$ | Entraining velocity at pitch line |
| $V_r$ | Viscosity ratio |
| $V_t$ | Viscosity ratio at operating temperature |
| $W$ | Load |
| $\hat{W}$ | Dimensionless load parameter |
| $\bar{W}$ | $W/(\eta U_0 B(L/h)^2)$ (Chapter 5) |
| $\bar{W}$ | Dimensionless load coefficient |
| $W_0$ | Load/unit length |
| $x$ | Coordinate in direction of rolling |
| $x_i$ | Distance from inlet to edge of Hertzian contact |
| $x_m$ | Coordinate for maximum pressure |
| $Y$ | Load factor |
| $y$ | Transverse coordinate |
| $z$ | Coordinate normal to plane of contact |
| $\alpha$ | Parameter defining accuracy of Hertz equations (Chapter 1) |
| $\alpha$ | Constant (Chapter 2) |
| $\alpha$ | Pressure viscosity coefficient (Chapter 7) |
| $\beta$ | Radius of asperity tip (Chapter 1) |
| $\beta$ | Mechanical constant defining load capacity (Chapter 3) |
| $\gamma$ | Surface energy (Chapter 2) |
| $\gamma$ | $R_y/R_x$ (Chapter 7) |
| $\Delta T$ | Temperature rise |
| $\delta$ | Normal approach of surfaces (Chapter 1) |

| | |
|---|---|
| $\delta_e$ | Normal approach at limit of elastic deformation |
| $\delta$ | Temperature–viscosity coefficient |
| $\varepsilon$ | Hysteresis loss coefficient (Chapter 2) |
| $\varepsilon$ | Eccentricity ratio |
| $\eta$ | Number of asperities per unit area (Chapter 1) |
| $\eta$ | Dynamic viscosity (Chapters 5, 6, 7, 8, 9) |
| $\eta_e$ | Effective dynamic viscosity |
| $\eta_f$ | Dynamic viscosity of oil in supply groove |
| $\eta_0$ | Viscosity at ambient temperature and pressure |
| $\eta_s$ | Viscosity at surface temperature, ambient pressure |
| $\eta_x$ | Viscosity at surface temperature, actual pressure |
| $\theta$ | Semiangle of conical asperity (Chapter 2) |
| $\theta$ | Polar coordinate (Chapter 6) |
| $\Lambda$ | Parameter defining amount of asperity contact (Chapter 2) |
| $\lambda$ | Parameter governing normal approach (Chapter 1) |
| $\lambda$ | Coefficient of rolling resistance (Chapter 2) |
| $\lambda$ | Specific film thickness ratio ($= h/R$) (Chapters 7, 8, 9) |
| $\mu$ | Coefficient of friction |
| $v$ | Poisson's ratio (Chapters 1, 7) |
| $v$ | Kinematic viscosity (Chapters 6, 7, 8, 9) |
| $\xi$ | $a\rho cV/2\alpha$ |
| $\rho$ | Density |
| $\sigma$ | Standard deviation |
| $\sigma_r$ | Composite roughness |
| $\sigma_Y$ | Uniaxial yield stress |
| $\tau$ | Shear stress |
| $\tau_S$ | Interfacial shear strength |
| $\psi$ | Plasticity index $\left( = \dfrac{E}{H}\left(\dfrac{\sigma}{\beta}\right)^{1/2}\right)$ (Chapter 2) |
| $\psi$ | Attitude angle (Chapter 6) |
| $\phi$ | $(1 + 2\gamma)^{-1}$ (Chapter 7) |
| $\phi$ | Normal pressure angle of gear teeth |
| $\phi(z)$ | Surface height distribution function |

*Subscripts*

| | |
|---|---|
| 1, 2 | Refer to the two surfaces of the contact |

# Chapter 1
# Surface Properties and Surface Contact

## 1.1 INTRODUCTION

Tribology is defined as 'the science and technology of interacting surfaces in relative motion and of related subjects and practices'. Therefore, if we are to understand tribological processes, we must understand how surfaces interact when they are loaded together, and it is the primary purpose of this chapter to develop such an understanding.

However, before we can consider surface interactions, we need to have a clear idea of what we mean by 'surface', and the meaning in the context of tribology is by no means as simple as we might expect. Traditionally, we might think of the surface of a solid body as simply the geometrical boundary between the solid and its environment: i.e. a surface in space which has molecules of the solid to one side of it and molecules of the environment to the other. However, such a definition is far too limited for our purposes. Tribological behaviour is influenced by the physical, chemical and mechanical properties of not only the surface material, but also the near-surface material; and when we talk to surfaces, we shall implicitly include the material to a significant depth below the actual boundary.

It will become clear in the remainder of this chapter and in the following chapters that the interactions between two surfaces are very dependent not only on the materials of the two contacting surfaces, but also on their shapes, and we therefore need an appropriate method of describing and measuring surface shape.

Thus, we start this chapter by describing the nature of the surface and near-surface regions of a solid; we shall then show how we describe and measure the shape of a surface; and, finally, we shall briefly describe those aspects of the mechanical interaction of surfaces which underlie our current understanding of tribological behaviour.

Primarily because of their high strength and toughness, metals have been the most common tribological materials for many years. For this reason, their tribological behaviour has been studied in greater detail, and is consequently more fully understood, than that of other materials. Therefore, we shall start by concentrating on the nature of the metal surface, but in later chapters we shall extend all our discussions to include more recent tribological materials such as ceramics, elastomers and polymers.

## 1.2 NATURE OF METALLIC SURFACES

A surface of a typical metal might appear as shown in Figure 1.1. On top of the normal crystalline structure of the bulk material lies a layer of deformed material resulting from the processes used in the manufacture of the surface. This deformed layer is itself covered by a compound layer resulting from chemical reaction of the metal with its environment. (With metals in air, this reacted layer will usually be an oxide, and, as we shall see, such oxide layers can critically affect both the friction and the wear of metals.) This reacted layer will, in a normal environment, often be covered with a layer of contaminants such as condensed oil vapours and particles of smoke and dust. Finally, machining processes may also cause contaminants such as cutting lubricants to be trapped in heavily deformed regions of the surface.

It should now be obvious that a metal 'surface' in a normal environment is a very complex structure containing varied amounts of different materials with very different properties, and we are unlikely to be able to make accurate predictions of the ways in which two such surfaces will interact when loaded together. It may, therefore, be thought that we should take as our baseline two completely clean surfaces – i.e. two surfaces from which all contaminants and reacted chemical films have been removed. Such surfaces can be generated in ultra-high-vacuum conditions, either in highly specialised laboratory apparatus or in outer space, and, as we shall see in the next chapter, the study of such surfaces has helped to elucidate some tribological processes.

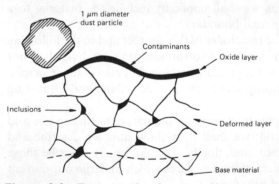

**Figure 1.1** Typical surface layers

However, apart from this particular case, other surfaces which we describe as clean will still retain the oxide films which are of paramount importance in determining tribological behaviour.

There are now many techniques available for studying the chemical nature of the surface and near-surface layers of materials. However, for the purposes of this text it is only necessary to make use of the relevant results of such studies: it is unnecessary to describe the actual techniques, and readers interested in these should consult specialised texts such as Walls (1989) and Woodruff and Delchar (1988).

## 1.3   SURFACE GEOMETRY

The geometric shapes of ordinary surfaces are controlled by the characteristics of the finishing processes by which they are produced. Close analysis of these surfaces shows that, even after the most careful polishing, they are still rough on a microscopic scale. Statistical analysis of the undulations of a surface normally shows that a very wide range of wavelengths is present, ranging from a fraction of a micrometer to many millimetres. Indeed, it is likely that the range of wavelengths detected is limited only by the resolution of the measuring instrument and that the wavelength spectrum actually extends from atomic dimensions to the full length of the surface being measured.

It has been common practice to describe different wavelength bands in different terms, referring to wavelengths of the order of micrometres as roughness and the longer wavelengths as waviness. However, such a sub-division is quite arbitrary, and the only practical necessity is to be aware of which undulations are functionally important in any particular application. In tribology the individual points of contact between two solid surfaces have dimensions of the order of micrometres, and we are, therefore, concerned with the amplitudes of wavelengths of the same order – i.e. with those wavelengths which have been conventionally described as roughness. The peaks of such surface undulations are known as asperities, and friction and wear both arise from the contacts between such asperities on opposing surfaces.

It is reasonable to assume that both friction and wear will depend in some way on the stresses arising at such asperity contacts and that the stresses themselves will depend on the heights of the asperities above the general level of the surface; in a later section we shall show more rigorously that this is indeed the case. We can, therefore, see that the tribological behaviour of two extended surfaces in contact will depend on the statistical distributions of asperity heights relative to the general level of the surfaces, and if we are to understand such behaviour, we must be able to measure and describe these distributions. The ways in which we do this form the subject of the following section.

## 1.4 SURFACE TEXTURE ANALYSIS

A typical surface might have $\simeq 10^5$ peaks. The problems of measuring the height and location of each of these peaks on each of two surfaces would be formidable. The problem of then analysing the interactions between the two sets of peaks when the surfaces are loaded together under normal and tangential loads would be overwhelming.

Any realistic approach to this problem must, therefore, use measurements from a small but representative sample of the surface – i.e. a sample chosen to be of such a size that there is a high probability that the surface lying outside the sample is statistically similar to that lying within it. In the remainder of this section we describe, first, the instrument used to obtain such measurements and, second, the ways in which the measurements are interpreted to give us descriptive parameters for the surfaces.

### 1.4.1 Surface Profilometry

The requirement for measuring a statistically representative sample of the surface is partly met by using a profilometer to sample a representative length of the surface with a high resolution in the plane normal to it. In the profilometer a very fine diamond stylus, with a tip radius of 2 $\mu$m or less, is drawn over the surface, as shown in Figure 1.2(a). The vertical movement of the stylus as it traverses the surface is then amplified and recorded to give

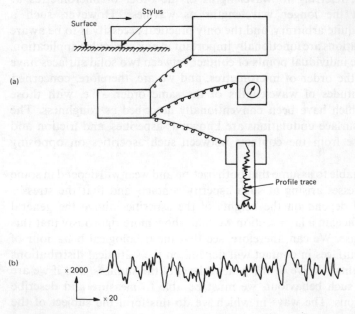

**Figure 1.2** (a) Schematic view of profilometer; (b) typical surface profile

a surface profile, as shown in Figure 1.2(b). In examining such profiles, some limitations of the method must be borne in mind, and these are discussed below.

The horizontal magnification is simply the ratio between the speed of the recording paper and the traverse speed of the stylus, and is typically $\times 100$. The vertical magnification is controlled electronically, and can normally be varied from $\times 500$ to $\times 100\,000$, according to the resolution required. This difference between the two magnifications is useful in giving greater emphasis to the important height characteristics of the surface, but it does mean that the resulting record is distorted. In Figure 1.3 a typical record is shown together with a record from the same surface using equal horizontal and vertical magnifications. The latter record shows that the asperities are undulations with slopes of a few degrees rather than the sharp peaks shown in the distorted profile.

Figure 1.3 also shows the size of the stylus to the same scale as the record, and it can be seen that some recording errors must occur due to the finite size of the stylus: the valleys will be recorded as being slightly narrower than their actual size, while the peaks wll be recorded as slightly broader.

The main disadvantage of profilometers of the type described above is that they are restricted to a single-line sample which may be unrepresentative of the whole surface if the texture has characteristics which are dependent on the orientation of the record. To overcome this disadvantage, there are now instruments available which can produce a three-dimensional map of the surface by recording a series of parallel and closely spaced profiles. However, for any surface which is approximately isotropic, the simpler instrument can provide measurements of the various parameters which are necessary for the characterisation of the surface.

It is also necessary to appreciate the difference between a peak on a profilometer trace and a true summit on the surface. The profilometer will register a peak each time it traverses a shoulder of a summit, and the number of peaks recorded will, therefore, be much greater than the number of true summits. This is clearly illustrated by the three-dimensional records of the type described above. This limitation is mitigated by the fact that any asperity on an opposing sliding surface will follow a similar path to that of the traversing stylus and the undulations over which it will have to pass are quite well represented by the profilometer trace.

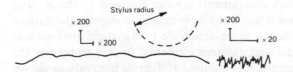

**Figure 1.3** Illustration of the profile distortion due to unequal vertical and horizontal magnifications

Finally, in profilometry the profile height must be measured with respect to some datum. The two most common methods of representing such data are the use of datum-generating attachments, which ensure accurate horizontal movement of the stylus support system, and the use of large-radius skids or flat shoes which rest on and traverse the surface being measured, and thus generate the general level of the surface. Each of these techniques creates some experimental errors, although the former is rather more precise.

In spite of the limitations discussed above, the profilometer remains the outstanding instrument for studying surface geometry and for evaluating its descriptive parameters. These parameters and the ways in which they are evaluated are described in the following sections.

### 1.4.2 Traditional Roughness Parameters

Until recent years, the analogue output of the profilometer was usually analysed within the instrument to give a graphical record of the profile and a single-parameter description of the surface roughness, which was displayed on an analogue meter. The two parameters in most common use were: (1) the roughness average $(R_A)$ value, defined as the mean vertical deviation of the profile from the centre line, treating deviations both above and below the centre line as positive; (2) the root mean square (RMS) value, defined as the square root of the mean of the square of these deviations. The $R_A$ has traditionally been the most common parameter in use in the UK, while the RMS value has been the most common in the USA.

These parameters are seen to be concerned only with relative departures from the centre line in the vertical direction; they do not provide any information about the shapes, slopes and sizes of the asperities or about the frequencies of their occurrence. It is, therefore, possible for surfaces of widely different profiles to give the same $R_A$ and RMS values, as shown in Figure 1.4.

These single-parameter descriptions are, therefore, mainly useful for comparing surfaces which have been produced to different standards but by similar methods: for example, lapped surfaces, differing only in the grade of lapping compound with which they have been finished.

### 1.4.3 Statistical Analysis of Surface Profiles

Surface profiles often reveal both periodic and random components in their geometric variation, and such components are not revealed by the $R_A$ and RMS values. In recent years it has become common to digitise the surface profile, to give a record which is more amenable to statistical analysis and to separation of its periodic and random elements. In this technique the analogue output from the amplifier is digitised at discrete intervals, as shown in Figure 1.5, to give a sequence of height readings relative to the profile centre line. A typical record would consist of 4000 digital height readings

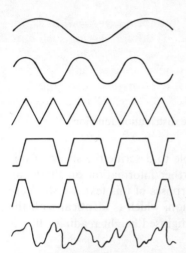

**Figure 1.4**   Various surfaces with the same $R_A$ value

taken at intervals of 2 $\mu$m along the surface. The sequence can then be analysed by an inbuilt microcomputer, to extract any chosen statistical information. The two most common types of information to be extracted are the height distribution of the surface texture and the amplitudes and wavelengths of periodic variations along the surface; we shall consider these in turn.

## The Height Distribution of the Surface Texture

As shown in Figure 1.5, the profile is sampled as height readings $z_1$, $z_2$, etc., at some discrete interval $l$. The resolution in the vertical direction is not infinite, and each height reading, $z$, will actually fall into one of perhaps 200 bands. The whole profile is then scanned by the computer, to determine the number, $N_z$, of ordinates detected in the band of mean height $z$, and a histogram of these occurrences can then be plotted, as shown in Figure 1.6. The profile can then be described by its all-ordinate distribution curve, $\phi(z)$, which is the best-fit smooth curve drawn through the histogram, and this distribution, in turn, can be converted into a probability distribution by normalising the readings so that the total area enclosed by the distribution curve is unity.

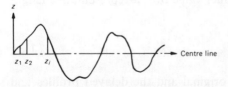

**Figure 1.5**   Surface height readings taken at discrete intervals

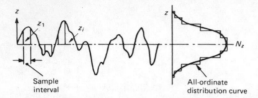

**Figure 1.6**  Method of deriving the all-ordinate distribution function

Many statistical parameters are available to describe the shape of this probability curve, and readers requiring further information on these can consult Halling (1975). However, for the purposes of this text, we shall use only the standard deviation of the distribution, which is identical with the RMS value defined earlier. In terms of the digitised height readings, this is defined by

$$\sigma = \text{RMS} = \left[ \frac{1}{n} \sum_{i=1}^{n} (z_i)^2 \right]^{\frac{1}{2}} \tag{1.1}$$

Note that we can also calculate the $R_A$ value from the statistical data, as it is simply given by

$$R_A = \frac{1}{n} \sum_{i=1}^{n} |z_i| \tag{1.2}$$

Other parameters have been used in other countries, and in recent years some of these, together with the RMS and $R_A$ values, have been precisely defined by the British Standards Institution in BS 1134. However, although several of these parameters are still widely used to specify required standards of surface finish on manufactured items, they are not as amenable to mathematical manipulation as is the $\sigma$ value, and in the following chapters we shall use this single parameter to characterise height variations.

### Periodic Variations along the Surface

Periodic height variations along the surface can be displayed by plotting the variation of the autocorrelation function $R(l)$ of the statistical data against the sampling interval $l$. $R(l)$ is obtained by delaying the profile relative to itself by some fixed interval $l$, multiplying the delayed profile by the original one, and averaging the resulting product value over a representative length of the profile. Thus,

$$R(l) = \frac{1}{N-l} \sum_{x=1}^{N-l} z(x) \cdot z(x+l) \tag{1.3}$$

where $l$ is the distance between the original and the delayed profiles and $N$ is the total number of ordinates in the total sample length $L$.

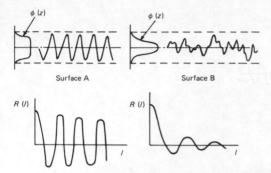

**Figure 1.7** Typical surfaces and the resulting autocorrelation functions

For a continuous profile, Equation (1.3) can be rewritten as

$$R(l) = \lim_{L \to \infty} \frac{1}{L} \int_{-L/2}^{L/2} z(x) \cdot z(x + l) \, dx \qquad (1.4)$$

It can then be seen that when $l = 0$, $R(l)$ reduces to the square of the $\sigma$ value of the profile. The autocorrelation function is therefore usually plotted in its standardised form $r(l)$, where

$$r(l) = \frac{R(l)}{\sigma^2} \qquad (1.5)$$

and the $r(l)$ value always has its maximum value of unity when $l = 0$.

Typical plots of autocorrelation functions for two different surfaces are shown in Figure 1.7. The shapes of these plots reveal both the random and the periodic characteristics of the profiles. The general decay indicates a decrease in correlation as $l$ increases, and is an indication of the random component, while the oscillatory component of the function reflects a similar periodicity in the profile. Figure 1.8 shows results from some machined surfaces, where the plots can be seen to demonstrate the general features of the profiles.

It can thus be seen that it is possible to describe the general features of any surface profile by two characteristics: the height distribution function $\phi(z)$ and the standardised autocorrelation function $r(l)$.

When we come to discuss the nature of contact between two surfaces, we shall also see that the heights, curvatures and frequencies of distribution of the surface asperities are of critical importance. All these data can be extracted from the digitised profile.

In practice, it is often found that both the all-ordinate distribution and the peak height distribution for engineering surfaces are good approximations to Gaussian distribution curves – i.e. curves of the form

$$\phi = \frac{1}{\sigma(2\pi)^{1/2}} \cdot \exp(-z^2/2\sigma^2) \qquad (1.6)$$

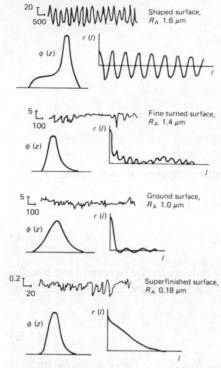

**Figure 1.8**  Examples of engineering surfaces, their distributions and their auto-correlation functions

This is by no means universally true, as full distributions often depart markedly from the Gaussian shape. However, surface contact normally involves no more than the outer 10% of the asperities, and even when the full distribution departs radically from Gaussian, this outer tail of the distribution is often a good approximation to the tail of a Gaussian distribution. This is a very important point, as we shall show later that it is this Gaussian or near-Gaussian distribution of asperities in the regions of contact which leads to the observed frictional behaviour.

## 1.5  SURFACE CONTACT

Both friction and wear are due to the forces which arise from the contact of solid bodies in relative motion. Therefore, any understanding of friction or wear must be based on an understanding of the mechanics of contact of solid bodies.

Solid bodies subjected to an increasing load deform elastically until the stress reaches a limiting value, known as the yield stress $\sigma_Y$, and at higher stresses they then deform plastically. In most contact situations some asperities are deformed elastically, while others are deformed plastically: the loads will induce a generally elastic deformation of the solid bodies but at the tips of the asperities, where the actual contact occurs, local plastic deformation may take place.

In this section we shall start by considering the elastic deformation of cylinders and spheres. This study is valuable for two reasons: first, many engineering situations involve the non-conforming contact of bodies defined by smooth curves; second, all solid bodies have surface asperities which can be considered as small spherically shaped protuberances, so that the contact of two macroscopically flat bodies can be reduced to the study of an array of spherical contacts deforming at their tips.

It is clear that a detailed study of surface contacts would involve a detailed understanding of the elastic and plastic deformation of contacting surfaces. However, it is the purpose of this chapter simply to set out the basic knowledge which we need to understand the material of the following chapters: it is not necessary fully to develop the theory, and readers requiring more detail are referred to the definitive book by Johnson (1985).

We shall be concerned with effects within the outermost layers of bodies, typically within a millimetre of the surface, and can ignore effects at greater depths. This allows us to treat the surfaces as surfaces of semi-infinite bodies, ignoring overall shape, and leads to mathematical simplification.

We are particularly interested in problems of contact between bodies whose geometries are defined by circular arcs. The problems of elastic contact between such bodies were first solved by Hertz in 1882, and such situations are referred to as Hertzian contact. We shall initially concentrate on the stresses in such contacts.

## 1.5.1 Stresses in Non-conforming Contacts between Curved Bodies

We consider first the contact of two identical cylinders under conditions of plane strain. From symmetry, we can see that compression of the cylinders will produce a plane-rectangular contact zone, as shown in Figure 1.9(a). (In the remainder of this chapter, to avoid ambiguity in use of the word 'contact', the shape of the contact zone will be referred to as the *footprint*.) For unequal cylinders, the footprint is no longer truly plane, but the discrepancy is negligible and plane contact can be assumed. As we increase the load, the width of the footprint will increase, and, as the deformation at the centre of the footprint is greater than that at its edges, we expect a contact pressure which varies with position. Therefore, to determine the stress distribution within the surface, we must define both the size of the footprint and the pressure distribution within it for any applied load. A detailed analytical treatment can be found in Johnson (1985), but we shall use a simple physical argument.

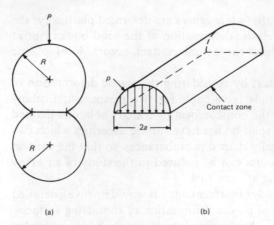

(a)                                    (b)

**Figure 1.9**  Pressure distribution due to contact of two cylinders

For two identical elastic cylinders in contact under a normal load, $P$, per unit length, let the resulting plane footprint have a width of $2a$, as shown in Figure 1.9(b). It is apparent that the stresses in such a system would be such that stress $\propto (P/a)$. The actual pressure distribution is given by

$$p = (2P/\pi a)(1 - x^2/a^2)^{\frac{1}{2}} \qquad (1.7)$$

We also note that an increase in load would cause increases in both the footprint half-width, $a$, and the strain, so we expect that strain $\propto (a/R)$, where $R$ is the radius of the cylinder. From these relations for stress and strain, we have $(P/a) \propto E(a/R)$ or $a^2 \propto PR/E$. The actual solution for this case is

$$a^2 = 4PR(1 - v^2)/\pi E \qquad (1.8)$$

The solution defined by Equations (1.7) and (1.8) is almost true for both identical and non-identical cylinders as long as the angle subtended by the contact width at the centre of the cylinder is less than 30°. It is also approximately true, over the same angular range, for contacts between different materials if we substitute $E^*$ and $R'$, respectively, for $E$ and $R$, where

$$1/E^* = (1 - v_1^2)/E_1 + (1 - v_2^2)/E_2$$

and

$$1/R' = 1/R_1 + 1/R_2$$

where $a^2 = 4PR'/\pi E^*$. For the case of a cylinder on a plane, the plane has infinite radius, so that $R'$ becomes the radius of the cylinder only; for concave curvatures the radius is negative.

We know that the onset of plastic deformation will be associated with the maximum shear stress in the material reaching a critical value, $k$, and we therefore wish to know the distribution of maximum shear stress for a body loaded with the pressure distribution given by Equation (1.7) acting over the footprint extending from $-a$ to $+a$.

The maximum shear stress in plane strain conditions is given by

$$\tau_{max} = \{(\sigma_x - \sigma_z)^2/4 + \tau_{xz}^2\}^{\frac{1}{2}} \tag{1.9}$$

where expressions for $\sigma_x$, $\sigma_z$ and $\tau_{xz}$ can be found in Johnson (1985). Therefore, Equation (1.9) defines the values of $\tau_{max}$ at all points. The evaluation of this expression allows us to plot the maximum shear stress distribution beneath the footprint, and the rather surprising result is shown by the isochromatics of the photoelastic stress pattern of Figure 1.10. The greatest value of $\tau_{max}$ occurs at a distance of $0.79a$ below the surface, and we find that it will attain the critical value of $k$ when the maximum pressure, $p_{max}$, at the centre of the footprint is $3.3k$. If we had simple compression, we should expect the surface material to yield when $p_{max}$ attained a value of $2k$, but this does not happen in the present case, because the surface elements are subjected to compressive stresses in all three orthogonal directions, giving a hydrostatic stress component which cannot contribute to plastic deformation.

This very important result means that higher loads than expected can be carried elastically in these non-conforming contacts. Furthermore, even when sub-surface yielding has taken place, very little plastic deformation can occur, because the plastic zone is constrained by the surrounding elastic material.

As the load is increased further, the plastic zone will increase in size and ultimately spread to the surface, so that plastic indentation can occur. This happens when the mean pressure is approximately $6k$ – i.e. approximately twice the pressure at which yield first occurred. The mean pressure at this point is essentially the indentation hardness value, $H$, of the material, which

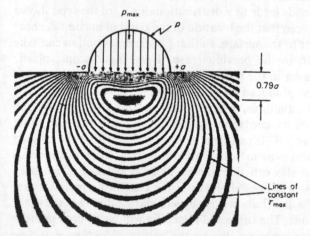

**Figure 1.10** Actual isochromatics obtained for the contact of a cylinder and a plane due to normal load alone

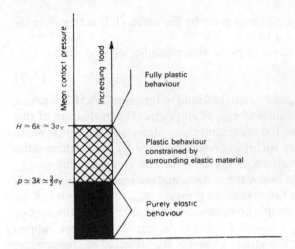

**Figure 1.11**

is why, for metals, we find that $H \simeq 6k \simeq 3\sigma_Y$, where $\sigma_Y$ is the uniaxial tensile yield strength of the material.

The behaviour described above is summarised in Figure 1.11.

We have, as yet, considered the effects of normal loads only, and to understand tribological phenomena, we must consider the effects of both normal and tangential loads. As we shall see in the next chapter, the tangential force, $F$, required to cause sliding under a normal load, $W$, is given by $\mu W$, where $\mu$ is a constant known as the coefficient of friction. The stress field due to a tangential force can be derived by a method similar to that described above for a normal force, and combining the stress distributions due to the normal and tangential loads leads to a distribution of $\tau_{max}$ of the type shown in Figure 1.12. It can be seen that the location of the point of maximum shear stress is now much closer to the surface, so that plastic deformation can take place more readily than in the previous case. Thus, macroscopic plastic deformation becomes easier in the presence of friction forces. A full solution of this problem has been produced by Hamilton (1983).

In many practical situations, where friction is used to prevent slip between mating components, contacting bodies are subjected to tangential forces which are less than $\mu W$. It can then be shown that within the contact area there exists a central area in which no slip occurs, while outside this area a small amount of slip can take place. This coexistence of zones of sticking and microslipping is possible because the contacting materials are deformable and the stress distribution is such as to allow slipping in the outer regions of the contact zone. The subject of microslip will be covered in more detail in the next chapter, when we discuss the nature of friction in rolling contacts.

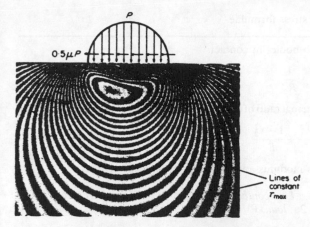

**Figure 1.12** Actual isochromatics obtained for the contact of a cylinder and a plane due to combined normal and tangential loads, where $T = 0.5\,\mu P$

For simplicity, we have concerned ourselves so far with two-dimensional non-conforming contacts, although in many practical situations we are concerned with three-dimensional problems. In such situations the patterns of behaviour are similar to those already described, but some of the equations describing the behaviour must be modified. For contact between two similar spheres we have a plane circular footprint and a hemispherical pressure distribution, as shown in Figure 1.13, given by

$$p = (3W/2\pi a^2)(1 - x^2/a^2 - y^2/a^2)^{\frac{1}{2}} \tag{1.10}$$

where the value of $a$ is given by

$$a = (3WR/8E^*)^{\frac{1}{3}} \tag{1.11}$$

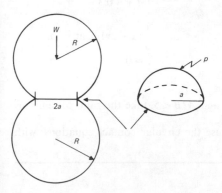

**Figure 1.13** Pressure distribution due to contact of spheres

Table 1.1  Elastic contact stress formulae

Suffixes 1, 2 refer to the two bodies in contact.

Effective curvature $\dfrac{1}{R} = \dfrac{1}{R_1} + \dfrac{1}{R_2}$

where $R_1$ and $R_2$ are the principal radii of curvature of the two bodies (convex positive).

Contact modulus $\dfrac{1}{E'} = \dfrac{1 - v_1^2}{E_1} + \dfrac{1 - v_2^2}{E_2}$

where $E$ and $v$ are Young's modulus and Poisson's ratio, respectively.

| | *Line contact*<br>(*width, 2a; load, P'/unit length*) | *Circular contact*<br>(*diameter, 2a; load, P*) |
|---|---|---|
| Semi-contact width or contact radius | $a = 2\left(\dfrac{P'R}{\pi E'}\right)^{\frac{1}{2}}$ | $a\left(\dfrac{3}{4}\dfrac{WR}{E'}\right)^{\frac{1}{3}}$ |
| Maximum contact pressure ('Hertz stress') | $p_0 = \left(\dfrac{P'E'}{\pi R}\right)^{\frac{1}{2}}$ | $p_0 = \dfrac{1}{\pi}\left(\dfrac{6WE'^2}{R^2}\right)^{\frac{1}{3}}$ |
| Approach of centres | $\delta = \dfrac{2P}{\pi}\left(\dfrac{1 - v_1^2}{E_1}\left(\ln\dfrac{4R_1}{a} + \dfrac{1}{2}\right) + \dfrac{1 - v_2^2}{E_2}\left(\ln\dfrac{4R_2}{a} - \dfrac{1}{2}\right)\right)$ | $\delta = \dfrac{a^2}{R} = \dfrac{1}{2}\left(\dfrac{9}{2}\dfrac{W^2}{E'^2R}\right)^{\frac{1}{3}}$ |
| Mean contact pressure | $P_{mean} = \dfrac{P'}{2a} = \dfrac{\pi}{2}p_0$ | $P_{mean} = \dfrac{P}{\pi a^2} = \dfrac{2}{3}P_0$ |
| Maximum shear stress | $\tau_{max} \approx 0.300\, p_0$ at $x = 0,\ z = 0.79a$ | $\tau_{max} \approx 0.310\, p_0$ at $r = 0,\ z = 0.48a$ for $v = 0.3$ |
| Maximum tensile stress | zero | $\dfrac{1}{3}(1 - 2v)p_0$ at $r = a$, $z = 0$ |

*Mildly elliptical contacts*

If the gap at zero load is $z = \frac{1}{2}Ax^2 + \frac{1}{2}By^2$ and $A/B < 5$, take the ratio of semiaxes $b/a \cong (A/B)^{\frac{1}{3}}$.

To find the contact area or Hertz stress, use the circular contact equations, with $R_e = (AB)^{-\frac{1}{2}}$.

Although the footprint for dissimilar spheres is not a plane circular area, Equation (1.10) still holds with substantial accuracy, but in this case we have

$$a = (3WR'/4E^*)^{\frac{1}{3}} \tag{1.12}$$

For contact between a sphere and a plane, $R'$ is simply the radius of the sphere.

For general contact between two curved surfaces, the footprint is an ellipse. This is the case in many important practical situations, particularly the contact between the balls and bearing races in rolling contact bearings. In such situations the equations become more complex; it is beyond the scope of this text to derive them or to state them in full, and readers requiring more information should consult either Johnson (1985) or Halling (1975). Fortunately, for mildly elliptical footprints, the equations for circular footprints can be used in a slightly modified form.

In Table 1.1 we summarise the most commonly required formulae for rectangular footprints, circular footprints and mildly elliptical footprints.

## 1.5.2   Surface Deformations in Non-conforming Contacts

In considering asperity contacts, where we approximate the asperity tip as a hemisphere, we shall find that it is necessary to know the normal approach of a sphere to a plane, and the resulting deformation, due to the application of a normal load.

Consider the contact between a sphere and a plane, as shown in Figure 1.14. The separation, $u$, of the surfaces at a distance $r$ from the centre of the circular footprint is given by

$$u = R - (R^2 - r^2)^{\frac{1}{2}}$$

which, when $r$ is small compared with $R$, is given by

$$u = (r^2/2R) \tag{1.13}$$

The normal approach is defined as the distance by which points on the two bodies remote from the deformation zone move together on application

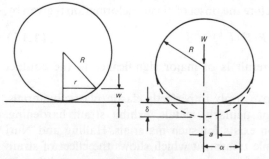

**Figure 1.14**   Elastic contact between a sphere and a plane

of a normal load; it arises from the flattening and displacement of the surface within the deformation region. If $a$ is the radius of the footprint and $w$ is the displacement of the sphere at the footprint boundary, then the normal approach, $\delta$, will be given by

$$\delta = u + w = (a^2/2R) + w \tag{1.14}$$

We know that at the centre of the footprint $\delta$ is the degree of deformation and it is reasonable to assume that the normal approach will be proportional to the flattening of the sphere – i.e.

$$\delta \propto a^2/2R$$

Substituting the value for $a$ given by Equation (1.12), we see that

$$\delta \propto (W^2/E^{*2})^{\frac{1}{3}}$$

The exact results show that

$$\delta = (9W^2/16E^{*2}R)^{\frac{1}{3}}$$

or

$$W = (4/3)E^*R^{\frac{1}{2}}\delta^{\frac{3}{2}} \tag{1.15}$$

Combining Equations (1.12) and (1.15), we see that the area of contact, $A$, will be given by

$$A = \pi a^2 = \pi R\delta \tag{1.16}$$

Equation (1.16) indicates that the surface outside the footprint is displaced in such a way that the area of contact is exactly half the area $2\pi R\delta$ which would be obtained by plastic flattening of the sphere.

Equation (1.16) also shows that for spherical contacts on a plane $\delta = a^2/R$.

As before, plasticity first occurs when the maximum elastic Hertzian contact pressure reaches $0.6H$, and we know that for this geometry $p_{max} = 1.5p_{mean}$, so that, at the onset of plasticity, $(3/2)W/\pi a^2 = 0.6H$. Substitution for $W$ and $\pi a^2$ from Equations (1.15) and (1.16) shows that the maximum normal approach before the onset of plastic deformation is given by

$$\delta_e = (0.3\pi)^2 R(H/E^*)^2 \simeq R(H/E^*)^2 \tag{1.17}$$

We shall see later that this result is of major significance in the contact of rough surfaces.

The results described above apply to perfectly elastic/plastic materials, while many real materials, particularly metals, exhibit strain hardening. Although no analytic solution exists for such materials, Halling and Nuri (1974) have developed a simple treatment which shows the effect of strain hardening. They consider the deformation behaviour of any material to be represented by a stress–strain law of the form

$$\sigma = \beta \varepsilon^n \tag{1.18}$$

When $n = 0$, we have rigid/perfectly plastic behaviour, where $\beta = \sigma_Y$; when $n = 1$, we have elastic behaviour, where $\beta = E$. They also assumed that the normal approach, $\delta$, could be defined by

$$a^2 = \lambda R \delta \tag{1.19}$$

with $\lambda$ having the values 1 when $n = 1$ and 2 when $n = 0$. They were able to make experimental measurements justifying these assumptions and, given the necessary experimental data, were thus able to define the normal approach and footprint size for the non-conforming contact of strain hardening materials.

### 1.5.3   The Effects of Surface Films

So far, we have considered contact situations involving, at the most, two different materials. However, tribological contacts often involve surfaces covered by films of other materials. These films may be adventitious – for example, due to chemical reaction with the environment; they may arise as a result of the rubbing process itself; or they may be deliberately formed using one of the processes described in Chapter 4. Clearly, as the film thickness tends to zero, the behaviour is determined entirely by the properties of the substrate, while, when the film thickness becomes very large, the behaviour is determined entirely by the properties of the film. Between these extremes the behaviour is a function of the properties of both the film and substrate materials. This is illustrated by the results of Figure 1.15, which show the

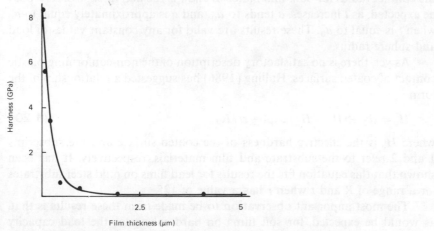

**Figure 1.15**  Microhardness values of lead films on steel substrates as a function of film thickness

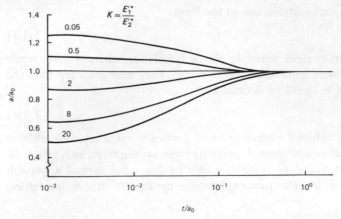

**Figure 1.16** Effect of film thickness on the Hertzian contact of coated surfaces

microhardness values for lead films of various thicknesses on mild steel substrates.

A study of the Hertzian contact of a sphere on a substrate coated with a different material (Sherbiney and Halling, 1976) gave the results shown in Figure 1.16. Here

$$1/E_1^* = (1 - v_1^2)/E_1 + (1 - v_3^2)/E_3$$

and

$$1/E_2^* = (1/v_2^2)/E_2 = (1 - v_3^2)/E_3$$

where the suffixes 1, 2 and 3 refer to the substrate, film and sphere materials, respectively. In this figure, $a$ is the actual footprint radius, $a_0$ is the value of the contact radius for bulk film material and $t$ is the film thickness. As would be expected, as $t$ increases, $a$ tends to $a_0$, and $a$ is approximately equal to $a_0$ when $t$ is equal to $a_0$. These results are valid for any constant values of load and sphere radius.

As yet, there is no satisfactory description of the non-conforming plastic contact of coated surfaces. Halling (1986) has suggested a relationship of the form

$$H_e = H_2 + (H_1 - H_2) \exp(-rt/R) \tag{1.20}$$

where $H_e$ is the effective hardness of the coated surface and the subscripts 1 and 2 refer to the substrate and film materials, respectively. It has been shown that this equation fits the results for lead films on mild steel substrates for a range of $R$ and $t$ when $r$ has a value of 125.

The most important observation to be made from these results is that, as would be expected, for soft films on hard substrates the load capacity decreases as film thickness increases, while for hard films on soft substrates the converse is true.

### 1.5.4   The Contact of Rough Surfaces

Although all surfaces are rough, we can simplify the study of two rough surfaces in contact if we consider the situation to be that of a single rough surface in contact with a perfectly smooth surface. We can further simplify the problem by modelling the asperities on the rough surface as sectors of spheres, so that their elastic deformation characteristics can be modelled by the Hertz theory for the contact between a sphere and a plane. Finally, we shall also assume that there is no interaction between the separate asperities, so that vertical displacement due to a load on one asperity does not affect the heights of the neighbouring asperities.

We consider first the simple, but unrealistic, model illustrated in Figure 1.17, where all the asperities on the plane of unit area have the same height, $z$, relative to the reference plane $XX'$. As the surfaces are loaded together, we see that the normal approach is $(z-d)$, where $d$ is the current separation of the smooth surface and the reference plane in the rough surface. Each asperity is deformed equally, and carries the same load, $W_i$, so that for $\eta$ asperities per unit area the total load $W$ will be $\eta W_i$. For each asperity, the load, $W_i$, and the area of contact, $A_i$, are known from the Hertzian theory. Thus, if $\beta$ is the asperity radius, we have

$$W_i = (4/3)E^*\beta^{\frac{1}{2}}(z-d)^{\frac{3}{2}}$$

and

$$A_i = \pi\beta(z-d)$$

The total load will then be given by

$$W = (4/3)E^*\beta^{\frac{1}{2}}(A_i/\pi\beta)^{\frac{3}{2}}$$

so that the load is related to the total area of contact, $A(=\eta A_i)$, by

$$W = (4E/3\pi^{\frac{1}{2}}\eta^{\frac{1}{2}}\beta)\cdot A^{\frac{3}{2}} \tag{1.21}$$

This result shows that, for this particular model, the real area of contact is proportional to the two-thirds power of the load when the deformation is elastic.

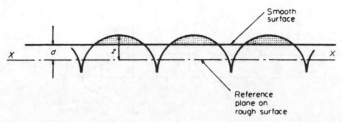

**Figure 1.17**   Contact between a smooth plane and an idealised rough surface

If the loads are such that the asperities are deforming plastically under the constant flow pressure, $H$, each individual contact area, $A'$, will be given by $2\pi\beta\delta$. The individual load, $W_i'$, will then be given by

$$W_i' = HA_i' = 2H\pi\beta(z - d)$$

Thus,

$$W' = \eta W_i' = \eta HA_i' = HA' = 2HA \tag{1.22}$$

so that the real area of contact is proportional to the load.

We shall see in Chapter 2 that it is basic to observed frictional behaviour that the real area of contact of rough surfaces should be proportional to the load. While our simple model gives this proportionality for plastic deformation, it does not do so for elastic deformation.

However, we know that our model of the rough surface is unrealistic. As pointed out earlier in this chapter, on real surfaces asperities have different heights following some probability distribution, and we must, therefore, modify our surface model to take this into account.

Therefore, we consider the contact of a rough surface, having a height distribution $\phi(z)$, with a smooth reference plane, as shown in Figure 1.18. If the separation of the two surfaces under load is $d$, then there will be contact of any asperity with an original height greater than $d$. The probability that any asperity has a height between $z$ and $(z + dz)$ above the reference plane will be $\phi(z)\,dz$. Thus, the probability of contact for any asperity of height $z$ is given by

$$\text{prob}(z > d) = \int_d^\infty \phi(z)\,dz \tag{1.23}$$

If we consider unit area of the surface containing $\eta$ asperities, the number of contacts, $n$, will be given by

$$n = \eta \int_d^\infty \phi(z)\,dz \tag{1.24}$$

Since the normal approach is $(z - d)$ for any asperity and $W_i$ and $A_i$ are known from Equations (1.15) and (1.16), the total area of contact and the

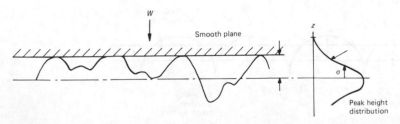

**Figure 1.18**  Contact between a smooth plane and a rough surface

corresponding load will be given by

$$A = \pi \eta \beta \int_d^\infty (z-d)\phi(z)\,dz \tag{1.25}$$

and

$$W = (4/3)\eta \beta^{\frac{1}{2}} E^* \int_d^\infty (z-d)^{\frac{3}{2}} \phi(z)\,dz \tag{1.26}$$

It is convenient and usual to express these equations in terms of standardised variables by putting $h = d/\sigma$ and $s = z/\sigma$, where $\sigma$ is the standard deviation of the peak height distribution of the surface. Thus,

$$n = \eta \int_h^\infty \phi^*(s)\,ds$$

$$A = \pi \eta \beta \sigma \int_h^\infty (s-h)\phi^*(s)\,ds$$

$$W = (4/3)\eta \beta^{\frac{1}{2}} \sigma^{\frac{3}{2}} E^* \int_h^\infty (s-h)^{\frac{3}{2}} \phi^*(s)\,ds$$

$\phi^*(s)$ being the probability density scaled to unit standard deviation.

Using these equations, the number of contact spots, the real area of contact and the load can be evaluated for any given height distribution.

An interesting situation arises for an exponential height distribution – i.e. $\phi(s) = e^{-s}$. In this case we have

$$n = \eta e^{-h}$$

$$A = \pi \eta \beta \sigma e^{-h}$$

$$W = \pi^{\frac{1}{2}} \eta \beta^{\frac{1}{2}} \sigma^{\frac{3}{2}} E^* e^{-h}$$

These equations give $W = C_1 A$ and $A = C_2 n$, where $C_1$ and $C_2$ are constants of the system. Thus, for this distribution, the area of contact and the number of contact spots are both proportional to the load, even though the asperities are deforming elastically.

For other distributions such a simple relationship will not apply, but it will be very nearly true for any distribution which approaches an exponential shape for the highest 10% of the asperities. As we stated in Section 1.4.3, many surfaces have distributions which are Gaussian or near-Gaussian in the outer 10% of the texture; with such surfaces the area of contact is found to be almost proportional to the load over several orders of magnitude of load.

## 1.5.5  Criterion of Deformation Mode

We shall see in the following two chapters that both friction and wear are very strongly influenced by the degree of plastic interaction of the contacting asperities. In most engineering situations some asperities will deform plastically, while others will deform elastically, and we thus have a mixed elastic–plastic contact where the greater the load and, hence, normal approach, the greater the number of plastic contacts. Thus, the normal approach should be an indicator of the degree of plasticity.

From Equations (1.15) and (1.16) we can see that the mean contact pressure, $p_{mean}$, for an elastic asperity contact is given by

$$p_{mean} = (4E^*\delta^{\frac{1}{2}})/(3\pi\beta^{\frac{1}{2}})$$

or

$$\delta^{\frac{1}{2}} = (3\pi\beta^{\frac{1}{2}}p_{mean})/4E^* \tag{1.27}$$

For a spherical contact we know that the transition from first yield to full plasticity takes place as the contact pressure increases from approximately $0.5H$ to $H$. To a good approximation, therefore, we can calculate the normal approach for the onset of plasticity by

$$\delta_e^{\frac{1}{2}} = (H/E^*)\beta^{\frac{1}{2}}$$

It is convenient to standardise this by dividing both sides by $\sigma^{\frac{1}{2}}$, so that

$$\delta^{*\frac{1}{2}} = (\delta_e/\sigma)^{\frac{1}{2}} = (H/E^*)(\beta/\sigma)^{\frac{1}{2}}$$

However, this parameter decreases as the surface roughness increases, and it is, therefore, usual to define a function $\psi$, *the plasticity index*, as the inverse of $\delta^{*\frac{1}{2}}$, so that

$$\psi = (E^*/H)\cdot(\sigma/\beta)^{\frac{1}{2}} \tag{1.28}$$

This index, first defined by Greenwood and Williamson (1966), is very indicative of the degree of plastic deformation. When the index is greater than unity, surface deformation is largely plastic, while when it is less than 0.6, the deformation is largely elastic; these statements are true *whatever the load*, as long as the asperities continue to deform independently.

The parameter depends on both the mechanical properties and the surface topography of the contacting surfaces; we shall return to this point when comparing the tribological properties of different types of materials in the following chapters.

Recently Halling *et al.* (1988) have extended this theory to allow predictions of the relative amounts of elastic and plastic deformation.

In the original formulation by Greenwood and Williamson (1966) the upper limit of the distribution curve was taken as infinity, as it must be, if the curve is truly Gaussian. However, this makes it impossible to define the surface separation at which contact first occurs (with a true Gaussian distribution, there is a finite, though very small, probability that first contact will occur when the mean separation of the surfaces is some metres). Halling *et al.* (1988) overcame this problem by truncating the distribution so that the probability of finding an asperity fell to zero at some finite distance, $b\sigma$, above the mean line. This is open to some reservations; in particular, the expected height of the highest asperity, as measured by $b$, will not have a constant value but will increase as the surface area increases. However, even for Gaussian surfaces, about 99% of all the asperities lie within a distance $3\sigma$ of the centre line and, as we shall see, truncation of the distribution at a value of $3\sigma$ – i.e. at a value of $b = 3$ – gives a very good fit of theory to experimental observations.

Halling *et al.* then considered the situation shown in Figure 1.19. It can be seen that the asperities from $d$ to $(d + \delta_e)$ are deforming elastically, while the asperities from $(d + \delta_e)$ to $b\sigma$ are deforming plastically. By integrating the distribution curve between these two sets of limits, it is then possible to define the ratio between plastically deforming area and elastically deforming area for different contact pressures and different values of the plasticity index, $\psi$. The results for truncation at $3\sigma$ are shown in Figure 1.20, where $\bar{R}$ is the ratio between plastic area and elastic area, and $\bar{p}$ is the non-dimensional nominal pressure $(N/AH)$. For $\bar{R} = 0$, we have the result for purely elastic deformation. It can be seen that this approach allows us to determine the relative amount of plastic deformation for different values of the plasticity index and different nominal pressures. As both friction and wear are very dependent on the degree of plastic deformation, it is likely that this approach will prove to be very useful in the future in helping us to predict friction and

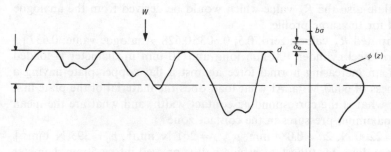

**Figure 1.19**  Elastic–plastic contact of a smooth surface and a rough plane with a peak height truncated at $b\sigma$

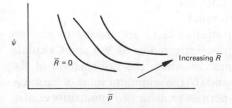

**Figure 1.20**  Ratio between plastic and elastic contact area as a function of plasticity index, $\psi$, and normalised pressure, $\bar{p}$

wear behaviour. The current difficulties in making such predictions will become abundantly clear in the following chapters.

Finally, it should be noted that with Hertzian contacts, such as bearings, gears and cams, the effect of roughness at low loads is to lead to contact sizes much larger than those predicted for smooth surfaces. A useful guide to the effect of roughness on Hertzian predictions has recently been proposed by Greenwood (1984). A parameter $\alpha$ is defined as

$$\alpha = \sigma' R' / a^2 \tag{1.29}$$

where $\sigma'$ is the combined roughness of the two surfaces, $R'$ is their relative radius of curvature and $a$ is the predicted Hertzian contact radius. Then, if $\alpha < 0.05$, the error in using the Hertzian equations to calculate the contact radius will be less than 7%.

## 1.6 PROBLEMS

(In solving these problems, the following properties may be assumed. Young's modulus(GPa): steel, 200; copper, 124; tungsten carbide, 550. Poisson's ratio: 0.3 in each case.)

1 A surface profile is sinusoidal, with unit amplitude and wavelength $\lambda$. The profile is sampled at equal intervals, with the origin on the centre line at a position of zero amplitude. Calculate the computed $R_A$ values for this profile for sampling intervals of $\lambda/2$, $\lambda/4$, $\lambda/8$ and $\lambda/16$. Calculate also the $R_A$ value which would be derived from the analogue signal for the same profile.
   [Computed $R_A$ values: zero; 0.5; 0.603; 0.628. Analogue value: 0.637]

2 A hard steel cylinder 100 mm long and 10 mm in diameter is loaded under an increasing normal force against a flat copper plate having a hardness of 6000 MPa. At what load does the material of the plate first yield; what is the corresponding contact width; and what are the mean and maximum pressures in the contact zone?
   [$P_Y = 2200$ N; $2a = 0.084$ mm; $p_{mean} = 261$ N/mm$^2$; $p_0 = 393$ N/mm$^2$]

3 A 2 mm diameter tungsten carbide ball is pressed into a 5 mm diameter hemispherical recess in a steel plate under a load of 50 N. Calculate (a) the minimum hardness of the steel plate for the contact to remain fully elastic; (b) the diameter of the contact zone.
   [(a) $H = 5245$ MPa; (b) $2a = 0.056$ mm]

4 Two similar, macroscopically flat steel surfaces are ground to give an RMS value of 0.2 $\mu$m and a mean asperity radius of 300 $\mu$m. Calculate (a) the maximum value of hardness at which the deformation of the asperities is predominantly plastic and (b) the minimum value of hardness at which the deformation of the asperities remains predominantly elastic. Take the mixed elastic–plastic range to be defined by $0.5 < \Psi < 1$.
   [(a) $H = 4750$ MPa; (b) $H = 7920$ MPa)]

# Chapter 2
# Friction

## 2.1 INTRODUCTION

Friction is the resistance to motion which occurs whenever one solid body slides over another. The resistive force, which acts in a direction directly opposite to the direction of motion, is known as the friction force. The friction force which is required to initiate sliding is known as the static friction force, while that required to maintain sliding is known as the kinetic friction force. Kinetic friction is usually lower than static friction.

### 2.1.1 The Laws of Friction

It has been observed experimentally that there are two basic 'laws' of friction, which are usually obeyed to a good approximation. The laws are entirely empirical and no physical principles are violated in those cases where the laws are not obeyed.

The first law states that the friction is independent of the apparent area of contact of the two bodies, and the second states that the friction force is proportional to the normal load between them. Thus, a brick can be slid as easily on its side as on its end, and if the load between two bodies is doubled, then the friction force is also doubled. These laws are known as Amonton's laws, after the French engineer who first presented them in 1699.

Coulomb in 1785 proposed a third law, that the kinetic friction is almost independent of the speed of sliding, but this law has a smaller range of applicability than have the first two.

### 2.1.2 The Coefficient of Friction

Amonton's second law states that the friction force, $F$, is proportional to the normal load, $W$ – i.e.

$$F = \mu W \tag{2.1}$$

27

where $\mu$, the constant of proportionality, is known as the coefficient of friction. As with the friction force, we can define coefficients of static and kinetic friction, but, unless stated otherwise, the coefficient of friction is normally taken to mean the kinetic coefficient.

It must be stressed that the coefficient of friction is not a constant for any particular material, but is typical of *two* materials, which may be similar or dissimilar, sliding against each other *under a given set of surface and environmental conditions*. The often-posed question 'What is the coefficient of friction of steel?' has no meaning.

### 2.1.3 The Basis of Amonton's Laws

To explain the form of Amonton's laws, we start by making the following two assumptions.

(1)  During sliding, the resistive force per unit area of contact is constant, so that

$$F = A \cdot s \tag{2.2}$$

where $F$ is the friction force, $A$ is the real area of contact and $s$ is the friction force per unit area.

(2)  The real area of contact, $A$, is proportional to the normal load, $W$ – i.e.

$$A = q \cdot W \tag{2.3}$$

where $q$ is the constant of proportionality.

Eliminating $A$ from these two equations gives

$$F = q \cdot s \cdot W \tag{2.4}$$

If we can justify our assumptions, then these equations explain the form of Amonton's laws: (2.2) and (2.3) tell us that the friction force depends upon the *real* area of contact and that this is independent of the apparent area, while (2.4) tells us that friction force is proportional to the normal load. It therefore remains to justify the assumptions.

Assumption (1) is easily justified; it involves no assumptions about the nature of the specific friction force, but is simply a statement that any part of the contact area is statistically representative of the whole.

Assumption (2) could not be justified under all circumstances, but we have already shown in Chapter 1 that it is justified: (a) whenever contact is wholly plastic, regardless of the surface topography; (b) whenever the contacting surfaces have exponential asperity height distributions, regardless of the mode of deformation; (c) to a very good approximation, whenever the contacting surfaces have a Gaussian asperity height distribution, again regardless of the mode of deformation.

Until the pioneering work of Greenwood and Williamson (1966), which demonstrated the critical importance of surface topography as summarised in (b) and (c) above, it was assumed that Amonton's second law implied that frictional contacts must be plastic. However, it is now clear, as explained in Chapter 1, that almost all rubbing surfaces will simultaneously contain both elastic and plastic contacts, and that it is the generally Gaussian or near-Gaussian surface topographies of most engineering surfaces which explain the wide range of validity of this law.

### 2.1.4   The Prediction of Friction Coefficients

The process of relative sliding of two materials appears to be a simple one, but this appearance is extremely deceptive. There is, as yet, no theoretical method of predicting the coefficient of friction (or, as we shall see in the next chapter, the rates of wear) when any two materials are in sliding contact. It should now be clear from a reading of Chapter 1 that both friction and wear are the results of extremely complex interactions between the surface and near-surface regions of the two materials. These regions differ from the bulk of the materials in their physical, chemical and mechanical properties. Furthermore, these properties themselves can change radically as a result of interactions of the surface atoms with their environments and with each other. It would be difficult to better the remark of Pauli, recently quoted by Tabor (1985), that 'God made solids, but surfaces were made by the Devil'.

As a result of this complexity, if it is necessary to know the coefficient of friction of a particular pair of materials under a particular set of operating conditions, then the safest procedure is to measure it experimentally, under conditions as close to the operating conditions as is feasible.

However, although accurate quantitative predictions of friction coefficients are not yet possible, the conceptual understanding of friction processes is now quite advanced, and this should help the practising tribologist to make sensible choices of pairs of materials for given rubbing conditions. It is the purpose of this chapter to impart this basic understanding. However, before turning to this, it will be useful to describe briefly how friction coefficients are measured.

### 2.2   THE MEASUREMENT OF FRICTION

In any apparatus for measuring friction, two specimens are placed together under a known normal load, one of the specimens is caused to slide relative to the other, and the tangential force resisting sliding is measured. There are many methods available, of varying degrees of complexity and cost.

Probably the simplest method is to use the inclined-plane test illustrated in Figure 2.1. In this test a plate of one of the chosen materials is securely

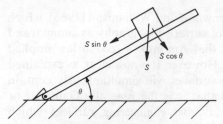

**Figure 2.1**  Measurement of friction using the inclined-plane test

fastened to a horizontal but tiltable plane surface; a block of the second material, of weight $S$, is placed upon it; and the plane is then gradually tilted until it reaches some angle $\theta$ to the horizontal at which the block begins to slide. At this point we have

$$\mu = F/W = S \sin \theta / S \cos \theta = \tan \theta$$

This method measures the coefficient of static friction and is obviously unsuitable in those cases where an investigation of the variation of friction with continuous rubbing is required. However, it can be used to make rapid measurements on many pairs of materials, and it is also a useful aid in teaching, where it can be used, for example, to demonstrate the validity of Amonton's laws.

Where continuous friction and wear measurements are required over a period of time, alternative tests must be used. In all such tests one specimen is driven continuously, while a second specimen, which is nominally stationary, is loaded against it. The loading of the stationary specimen can be by simple dead weight or, if the experimental conditions demand it, by some more complex mechanism such as spring loading, or hydraulic or pneumatic pressure. The measurement of the friction force is usually accomplished by mounting the nominally stationary specimen so that the friction force causes a very small tangential movement of either a capacitive or an inductive transducer, which is calibrated to give a continuous measurement of friction force.

The testing geometries for these more complex tests can be divided into two groups: conformal geometries, where the contact pressures are moderate and normally constant throughout the test, and non-conformal geometries, where the contact pressures are initially high and often vary with time. Examples of conformal tests and non-conformal tests are illustrated schematically in Figures 2.2 and 2.3, respectively. The conformal geometries can be used to simulate the tribological behaviour of devices such as brakes, thrust washers, plain bearings or face seals, whereas the non-conformal tests can be used either to simulate the behaviour of high-pressure contacts, such as those occurring between gear teeth, or, for example, to provide accelerated tests of the friction and wear behaviour of a number of candidate material pairs.

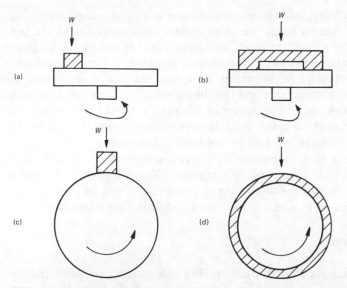

**Figure 2.2**  Measurements of friction using conformal geometries: (a) flat pin on disc; (b) thrust washer; (c) pin on cylinder; (d) shaft in bush

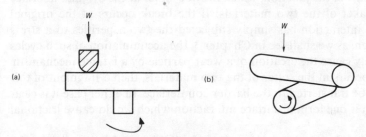

**Figure 2.3**  Measurements of friction using non-conformal geometries: (a) hemispherical pin on disc; (b) crossed cylinders

## 2.3  THE ORIGINS OF FRICTION

The purpose of this section is not to give a state-of-the-art review of the large amount of research that has taken place in this area, but rather to give a clear understanding of the concepts involved.

We know that the friction force is due to interactions between the opposing asperities of the two sliding surfaces. Each asperity interaction will contribute to the friction force, so that the total friction force at any time will be the sum of the forces at the individual contacts. Furthermore, as the surfaces move relative to one another, the energy supplied by the forces causing the motion is continuously dissipated at or near the sliding surfaces and, again, the energy dissipated in unit time will be the sum of the energies dissipated in the individual processes occurring during the same time. The observation that friction forces are often almost constant is due to the fact that the number of asperity interactions taking place at any time is so large that the statistical distribution of the contact processes is almost constant.

Clearly, in equilibrium the energy dissipated per unit time is equal to the product of the friction force and the sliding velocity, so we can approach the problem of explaining the nature of friction by considering either the origins of the forces at the contacts or the processes of energy dissipation. In what follows it will be convenient to use both approaches.

In considering asperity interactions, it is conventional to describe two types of interaction: adhesion and deformation. This convention will initially be followed here, but as the discussion is developed, it will be stressed that their contributions to friction are not simply additive, but interactive.

### 2.3.1   Adhesion

When two surfaces are loaded together, they can adhere over some part of the true contact area, to form friction junctions, and these junctions must then be broken if relative sliding is to take place. The break will occur at the weakest part of the junction, which may be either at the original interface or in the weaker of the two materials. If the break occurs at the original interface, the interaction has simply subjected the two asperities to a stress cycle, although, as we shall see in Chapter 3, the accumulation of such cycles may ultimately cause the creation of a wear particle by a fatigue mechanism. If the break occurs in the softer of the two materials, then a fragment of this material will be transferred to the harder counterface. In either case it is clear that adhesion is one form of surface interaction which would cause frictional resistance.

Before considering the contribution of adhesion to friction, it is useful to describe the adhesion of two solids placed in static contact. In doing this, use is made of the thermodynamic concepts of surface and interfacial energies.

The surface free energy of a solid is defined as the thermodynamically reversible energy involved in creating unit area of new surface. Similarly, if two surfaces are brought into contact, the interfacial free energy is defined as the thermodynamically reversible energy involved in creating unit area of interface. Thus, if we take two surfaces having surface energies $\gamma_1$ and $\gamma_2$ and place them in contact, to form an interface with interfacial energy $\gamma_{1,2}$, then the energy released is

$$\Delta\gamma = \gamma_1 + \gamma_2 - \gamma_{1,2} \tag{2.5}$$

This quantity is known as the thermodynamic work of adhesion as, in principle, a similar amount of work is needed to separate the bodies.

Although Equation (2.5) has been applied very successfully to liquids, it has been shown by Tabor (1985) that it cannot be applied precisely to even ideally elastic/brittle solids. For solids of other types, the energy calculated from Equation (2.5) is only a small fraction of the energy actually required to separate the surfaces; the reasons for this will be made clear later. However, although the thermodynamic work of adhesion is of little relevance in many contacts, the actual work and force needed to separate the surfaces are highly relevant. Tabor (1985) proposed for these the apposite terms 'pull-off work' and 'pull-off force' and these will be adopted here.

## 2.3.2 Deformation

If no adhesion takes place, then the only alternative interaction which would result in a resistance to motion would be one in which material is deformed and displaced during relative motion. We need consider only two interactions of this type: the microscopic interaction illustrated in Figure 2.4(a), where deformation and displacement of the interlocking surface asperities are required, and the more macroscopic interaction illustrated in Figure 2.4(b), where the asperities of the harder material plough grooves in the surface of the softer. Other examples of the second type of interaction would involve

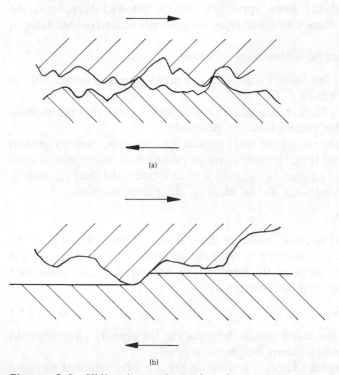

(a)

(b)

**Figure 2.4** Sliding interactions of surfaces: (a) asperity interaction only; (b) macroscopic ploughing interaction

the ploughing of one or both surfaces by wear particles trapped between them, and truly macroscopic ploughing of the softer material by the harder, with the dimensions of the ploughed groove being orders of magnitude greater than those of the asperities on either surface.

Asperity interactions will always be present and, together with adhesion, will often be the major cause of friction. The ploughing contribution may or may not be significant; its magnitude will depend on the surface roughnesses and relative hardnesses of the two surfaces, and on the size, shape and hardness of any wear debris and reaction products trapped between them. In the following section we shall describe how the three processes adhesion, ploughing and asperity interaction contribute to friction.

## 2.4  THEORIES OF FRICTION

### 2.4.1  Simple Adhesion Theory

This theory, due to Bowden and Tabor (1950), was the first modern explanation of the existence of friction. It was developed for ideal elastic–plastic metals and the same approach will be followed here, with the importance of the theory to other types of materials being covered later in the chapter.

The theory can be summarised as follows.

(1) When surfaces are loaded together, they actually make contact only at the tips of asperities.

(2) Even at low loads, the real contact pressures are so high that the asperity tips of the softer material deform plastically.

(3) This plastic flow causes the total contact area to grow, both by growth of the individual initial contacts and by initiation of new contacts, until the real area of contact is just sufficient to support the load elastically.

(4) Under these conditions, for an ideal elastic–plastic material

$$W = A \cdot p_0$$

where $W$ is the normal load, $A$ is the real area of contact and $p_0$ is the yield pressure of the softer of the two materials. The yield pressure is very nearly the same as the hardness, $H$, measured in an indentation test, so that this can be rewritten as

$$W = A \cdot H \tag{2.6}$$

(5) As a result of the severe plastic deformation, the asperity junctions cold weld – i.e. strong adhesive bonds are formed.

(6) The specific friction force, $s$, is then simply the force required to cause shear failure of unit area of asperity junction, so that

$$F = A \cdot s \tag{2.7}$$

Neglecting, for the moment, any ploughing contribution, Equations (2.6) and (2.7) can be combined to give

$$F/W = \mu = s/H \qquad (2.8)$$

It can be seen that the simple theory provided what was the first theoretical explanation of Amonton's laws: that friction is independent of the apparent area of contact and that friction force is proportional to the normal load.

In the above analysis we considered an ideal elastic–plastic material and ignored the effect of work hardening. For the case of failure in the softer of the two materials, we can take $s$ equal to $k$, the critical shear stress of this material, so that

$$\mu = k/H \qquad (2.9)$$

This ratio $k/H$ is fairly constant for most materials, and Equation (2.9) indicates why many material pairs have similar friction coefficients despite the fact that the individual values of $k$ and $H$ vary very widely. As shown in Chapter 1, a typical value of $k/H$ is $\simeq 0.16$, which is of the same order as the observed values of $\mu$ for many material pairs in normal atmospheres.

According to this simple theory, the effect of ploughing was taken into account simply by adding a term $\mu_{\text{ploughing}}$ to the coefficient calculated from Equation (2.9). The derivation of such ploughing terms is described in sub-section 2.4.3.

Although this simple theory is attractive, it is inadequate in several important respects.

(1) The friction coefficient given by Equation (2.9) depends only on the mechanical properties of the softer of the two materials, so we should expect a particular material to have the same friction coefficient against *any* harder counterface: this is not found to be the case.

(2) Actual values of $\mu$ in normal atmospheres, particularly for metal pairs, are normally of the order of 0.5, rather than the value of $\simeq 0.16$ which is given by Equation (2.9).

(3) Many materials, particularly ductile metals, exhibit friction coefficients much higher than those quoted above when their surfaces are free of the normal contaminant and oxide films – a condition which can be achieved in ultra-high vacuum. Friction coefficients much greater than unity are often observed, and in extreme cases gross sliding does not occur at all and a weld is formed over much of the nominal contact area.

These deficiencies led Bowden and Tabor to re-examine some of the assumptions of the simple adhesion theory, and to present a more realistic description of friction in terms of adhesion.

## 2.4.2 Extension of the Adhesion Theory

In the simple adhesion theory the effects of the normal and tangential loads were considered separately; the true area of contact was assumed to be determined by the normal load only (Equation 2.6) and the friction force was then taken to be the force required to cause shear over this area (Equation 2.7). However, the very high friction coefficients observed between clean metal surfaces indicate that the true area of contact must be far greater than that predicted by the simple adhesion theory. This was explained when Bowden and Tabor considered the combined effect of the normal and the shear stresses.

To illustrate this, we consider first the simple two-dimensional stress system shown in Figure 2.5(a), and assume that yielding occurs when the maximum shear stress attains the critical value $k$. At this point $k$ is given by the radius of the Mohr's circle shown in Figure 2.5(b). (The Mohr's circle construction is an illuminating technique for illustrating compound stresses, and readers who are not familiar with it should consult any standard text on strengths of materials, e.g. den Hartog, 1952). Using this construction it can be seen that yield occurs when

$$(p/2)^2 + \tau^2 = k^2$$

or

$$p^2 + 4\tau^2 = 4k^2 \qquad (2.10)$$

We can now examine the effects of the combined normal and shear stresses on the real area of contact in an asperity junction.

Consider first a single asperity junction under a normal load $W$, having an area of contact $A$, defined by $A = W/H$. Note that the material is at its yield point, defined by $H$, and that the maximum shear stress in the material is equal to $k$, so that any increase in either the normal stress or the shear stress in the junction will cause further plastic flow. Thus, if we impose an increment of tangential force, $f$, some plastic flow will occur and this will lead

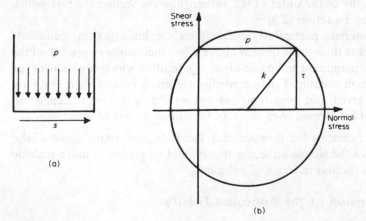

Figure 2.5 Mohr's circle construction for finding the maximum shear stress for an idealised two-dimensional junction under normal and tangential stress

to an increase in the contact area. This process is known as junction growth. As the area of the junction grows, both the normal and the shear stresses will decrease, and this process will continue until the maximum shear stress in the junction has fallen to its previous value of $k$, when the load will again be just supported elastically.

If we then impose further increments of tangential force, the above process will be repeated each time – i.e. the junction will continue to grow and no gross sliding will occur. In such cases the term 'coefficient of friction' is not strictly applicable, but, as the normal force remains constant while the tangential force continues to increase, the ratio between tangential stress and normal stress can attain a very high value, limited only by the gross shear strength of the softer of the two materials. Indeed, once strong adhesion has occurred, a strong resistance to shear can still be experienced if the normal load is removed, or even reversed.

We would expect the combined normal and shear stresses to obey an equation of a form similar to that given in Equation (2.10) for a simple two-dimensional junction. The exact form for a three-dimensional case is not known, but we assume that it is of the form

$$p^2 + \alpha \tau^2 = r^2 \qquad (2.11)$$

where $\alpha$ and $r$ are constants which we can determine by considering two boundary conditions.

(1) When there is no tangential force, $\tau$ is zero and $p = H$. Therefore, $r^2 = H^2$ and

$$p^2 + \alpha \tau^2 = H^2 \qquad (2.12)$$

(2) When large-scale junction growth has taken place, the friction force is much greater than the normal load, and, as these loads both act over the same area, the normal stress must be much less than the shear stress – i.e. $\tau \gg p$. Therefore, $\alpha \tau^2 \approx H^2$. In this case $\tau$ must be approximately equal to $k$, so that

$$\alpha \approx H^2/k^2 \qquad (2.13)$$

From Equation (2.13) we should expect $\alpha$ to have a value of approximately 36. However, experiments suggest that the value of $\alpha$ is lower than this, and Bowden and Tabor assumed a value of 9; as we shall see below, the exact value of $\alpha$ has only a secondary effect on the amount of junction growth which takes place.

Before moving on, it is useful to re-examine the second boundary condition described above – i.e. that when large-scale junction growth has taken place, the shear stress is very much greater than the normal stress. As the shear stress cannot exceed the shear strength of the softer material, this tells us that the normal pressure must now be very much lower than the yield pressure of that material. This is clearly the case, as the normal load has remained constant, while the area over which it acts has greatly increased, but it is worth emphasising that junction growth continues under the increasing shear stress, even when the normal stress is negligible.

This revised treatment was still based on the assumption that all junctions initially deform plastically under the applied normal load. However, this is an unnecessary restriction, as the same type of reasoning can also be applied to the many asperity contacts which, as shown in Chapter 1, deform elastically under the applied normal load. In such contacts, although the initial contact may be well within the elastic range, the same shear stress will still be required to shear the junction, and for clean surfaces this will be the full shear strength of either the interface or the weaker of the two materials. Thus, a contact which was initially elastic will be subjected to plastic shear at a low level of normal stress – a condition which is similar in all respects to the shearing of an initially plastic contact after some junction growth.

We have now explained the very high friction coefficients which can be obtained with clean metals in vacuum. We have not yet explained why such high coefficients are not observed in normal atmospheres, but it is clear that junctions formed in such atmospheres must be weaker in shear than those formed between clean surfaces, and that this relative weakness must be due to the presence of the contaminant films which are always present in normal atmospheres. Following Bowden and Tabor, we therefore assume that at any junction there is a thin contaminant film of shear strength $\tau_s$, and that $\tau_s = ck$, where $c$ is less than unity. While the frictional stress, $F/A$, is less than $\tau_s$, junction growth will proceed as described above; but when $F/A = \tau_s$, the contaminating film will shear, junction growth will end and gross sliding will occur.

Thus, the condition for sliding is

$$p^2 + \alpha\tau_s^2 = H^2$$

But it has already been shown that

$$H^2 = \alpha k^2$$

Therefore,

$$p^2 + \tau_s\alpha^2 = \alpha k^2$$

or

$$p^2 + \alpha\tau_s^2 = \alpha\tau_s^2/c^2$$

Hence,

$$\mu = \tau_s/p = \frac{c}{[\alpha(1-c^2)]^{\frac{1}{2}}} \tag{2.14}$$

According to this theory, as $c$ tends to unity, then $\mu$ tends to infinity, in agreement with the previous result for clean metals. The friction coefficient, $\mu$, is plotted against $c$ for various values of $\alpha$ in Figure 2.6, and we see that $\mu$ falls rapidly as $c$ reduces from unity. Thus, a small amount of weakening at the interface produces a drastic reduction in $\mu$. It is also clear from Figure 2.6 that, as mentioned earlier, the exact value of $\alpha$ is not of major importance.

Note that when $c$ is small,

$$\mu \simeq c/\alpha^{\frac{1}{2}}$$

so that we can obtain a low friction coefficient by arranging to have a film of low shear strength separating the surfaces. This is the principle underlying lubrication by soft metal films and boundary lubricants, which will be discussed later.

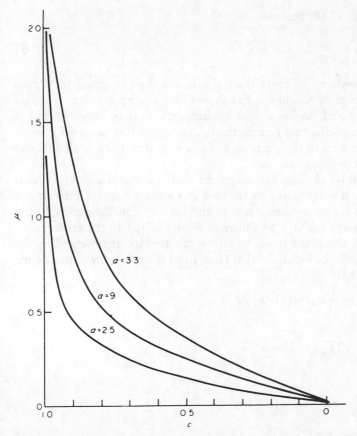

**Figure 2.6** Variation of $\mu$ with $c$ for different values of $\alpha$. It can be seen that, except at large values of $c$, the exact value of $\alpha$ is not of major importance

It is worth noting that this modified theory will also apply when the interface is inherently weaker than either of the asperity materials, even in the absence of contaminating films: in this case $\tau_s$ is simply the shear strength of the interface.

In this final modification to the adhesion theory, unlike the case for clean surfaces, there can be a very marked difference in behaviour between the junctions which are initially plastic and those which are initially elastic. In the former case there will always be some continued plastic deformation and junction growth until the interface shears, whereas in the latter case the interface may shear while the deformation of the asperities is still within the elastic range. The importance of this fact will emerge in Chapters 3 and 4.

In the foregoing discussion we have presented the various versions of the adhesion theory as first put forward by Bowden and Tabor. Johnson (1968) has suggested that Equation (2.11) should be modified to

$$p^2 + \alpha_1 \tau_s = \alpha_2 k^2 \tag{2.15}$$

so that Equation (2.14) becomes

$$\mu = \frac{c}{[\alpha_2 - \alpha_1 c^2]^{\frac{1}{2}}} \tag{2.16}$$

Then, if $\alpha_2$ exceeds $\alpha_1$ by more than 1, $\mu$ is less than or equal to 1, even when $c$ has a value of 1. Johnson has shown that as $c$ approaches 1, sliding causes $(\alpha_2 - \alpha_1)$ and, hence, $\mu$ also to approach 1. It is then possible to produce friction coefficients greater than 1, not by further junction growth, but by internal shear in the junction to form a chip or prow with adhesive transfer to the opposing surface.

More recently, Halling has suggested that $\alpha$ is not a constant but is itself a function of $c$. He considers the case of a surface element subjected to a normal stress $p$ and a shear stress $\tau$, and kept in equilibrium by lateral internal stresses $a_1 p$ and $a_2 p$, as shown in Figure 2.7(a). In the general case, where $0 < c < 1$, the plane strain situation can then be represented by the Mohr's stress circle of Figure 2.7(b). Then, from the geometry of this figure, it can be seen that

$$a_2 p = 1/2(\sigma_1 + \sigma_2) = (1 + a_1) \cdot p/2$$

and

$$p^2 [(1 - a_1)/2]^2 + c^2 k^2 = k^2$$

Therefore,

$$p^2 + \alpha c^2 k^2 = \alpha k^2$$

where

$$\alpha = [2/(1 - a_1)]^2 \tag{2.17}$$

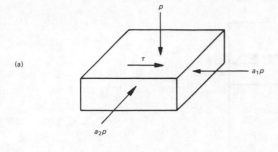

(a)

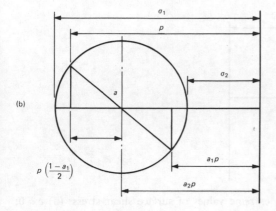

(b)

**Figure 2.7** Stresses on a surface element in plane strain

Hence,

$$\mu = ck/p = c/[\alpha(1 - c^2)]^{\frac{1}{2}} \tag{2.18}$$

It can be seen that Equation (2.18) is of the same form as Equation (2.14), but the constant $\alpha$ has been replaced by the variable given by Equation (2.17).

The values of $\alpha$ for the extreme cases of $c = 0$ and $c = 1$ can be calculated from the Mohr's circles shown in Figures 2.8(a) and 2.8(b), respectively. It can be seen, for a rigid–plastic material in plane strain, that when $\tau = 0$, $a_1 = a_2 = \frac{2}{3}$, while when $\tau = k$, $a_1 = a_2 = 1$. Inserting these values into Equation (2.17) shows that $\alpha$ has a minimum value of 36 when $c = 0$ and tends to infinity as $c$ tends to 1. The variation of $\alpha$ between these limiting conditions is not known, and the equations would need to be modified for the more realistic case of general triaxial stress and for either fully elastic deformation or plastic deformation with work hardening. However, this treatment does bring out the important point that a friction coefficient will depend on both the interfacial shear strength, as represented by $ck$, and the deformation characteristics of the material.

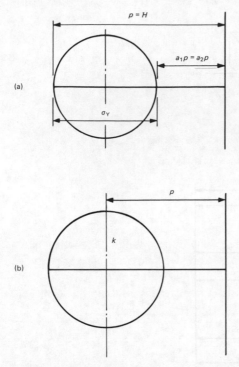

**Figure 2.8** Mohr's circles for extreme values of surface shear stress: (a) $c = 0$; (b) $c = 1$

### 2.4.3 Ploughing

The effect of a hard asperity ploughing a groove in a softer material is illustrated here for the example of a hard conical asperity of semiangle $\theta$, as shown in Figure 2.9. During rubbing, only the front surface of the asperity is in contact with the softer surface. Therefore, the normal load, $W_i$, which

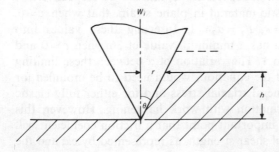

**Figure 2.9** Ploughing of a soft surface by a conical asperity

is supported by the horizontal projection of the asperity contact, is given by

$$W_i = \tfrac{1}{2}\pi r^2 \cdot H$$

and the friction force, $F_i$, which is supported by the vertical projection, is given by

$$F_i = r \cdot h \cdot H$$

Therefore,

$$\mu = F_i/W_i = 2h/\pi r = 2\cot\theta/\pi \tag{2.19}$$

Similar expressions can be calculated for other asperity shapes. For most engineering surfaces, the asperity angles are large and the ploughing component of friction due to asperity interaction is correspondingly small. For example, Equation (2.17) shows that for an asperity angle of 85°, which corresponds to an unusually rough surface, the ploughing coefficient of friction would be only 0.056. However, abrasive material, including work-hardened or oxidised wear debris may be very angular, and when significant amounts of such material are present between the rubbing surfaces, the ploughing component of friction may be much higher.

## 2.4.4  Deformation Theories

In the adhesion theory described above the normal and yield stresses on a single asperity were assumed to be representative of the stresses on all asperities. In deformation theories, it is recognised that the normal and shear stresses on the asperities will vary during the lifetime of a junction. The physical basis of the analysis is that, in the sliding of macroscopically flat surfaces, motion is parallel to the interface and the separation of the surfaces remains constant. This must be so, to maintain the area of contact at the constant level which will support the constant normal load, and the significant consequence of this is that the contacting asperities must deform to allow movement to continue.

This type of theory was first advanced by Green (1955) and has been extended by other workers. To illustrate the principles of such theories, the treatment of Edwards and Halling (1968) is summarised here.

Edwards and Halling considered two wedge-shaped asperities of semiangle $\theta$ moving as illustrated in Figure 2.10. By assuming that the solids are ideally elastic–plastic and that the asperities deform plastically, it is possible to calculate the instantaneous values of shearing force $F_i$ and normal load $W_i$ over the complete life of the junction. This was done both for the case of intimate contact, where the shear strength of the junction was taken to be $k$, and for the case of asperities separated by a film of shear strength $ck$, where $c$ and $k$ have the same meanings as in the Bowden and Tabor theory. Calculations were made for various values of $\theta$ and $c$. The variations

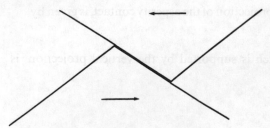

**Figure 2.10** The idealised wedge-shaped asperities studied in the 'plastic interaction theory'

of $F_i$ and $W_i$ with displacement are shown in Figure 2.11 for an asperity angle of 10° and two values of $c$. The coefficient of friction can then be computed as the instantaneous value of $\Sigma F_i / \Sigma W_i$, where the summation is made over all contacting asperities, or, alternatively, as the mean value of $F_i / W_i$ for a single pair of asperities over the full life of the junction. Calculated values of $\mu$ for various values of $c$ and $\theta$ are shown in Figure 2.12.

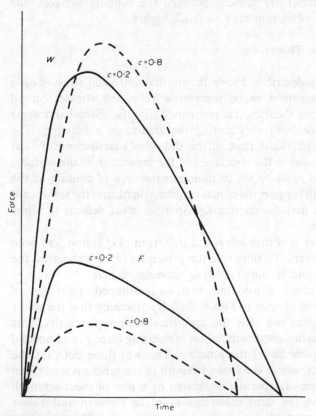

**Figure 2.11** Variation of the normal force, $P$, and the friction force, $F$, throughout the junction life for a junction of angle 10°

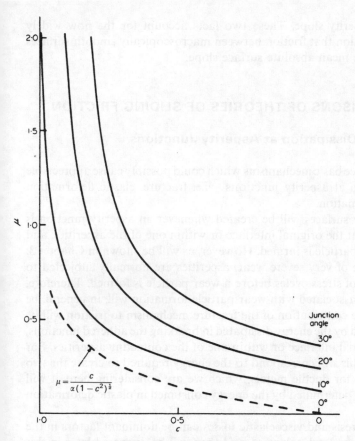

**Figure 2.12**   Variation of $\mu$ with $c$ for various junction angles

The general equation for $\mu$ given by the Edwards and Halling theory is

$$\mu = \left[ \frac{c}{\alpha(1-c^2)^{\frac{1}{2}}} + \phi \right] \bigg/ \left[ 1 - \frac{c}{\alpha(1-c^2)^{\frac{1}{2}}} \cdot \phi \right] \qquad (2.20)$$

where $\phi$ is a function of $c$ and the geometry of the junction, and is equal to zero when $\theta$ is zero.

It is interesting to note that when $\phi$ is zero, this equation reduces to a form which is identical with the modified Bowden and Tabor theory (Equation 2.14).

It is a characteristic of all asperity interaction theories of friction that the energy consumed in plastic deformation increases with the sharpness of the asperities. This is illustrated in Figure 2.12 for the Edwards and Halling model, and similar trends can be shown to occur with, for example, hemispherical asperities of decreasing curvature. In the case of elastic deformation of asperities, there is no deformation component of friction, but the real contact area of a junction and, hence, the force required to shear it increase with asperity slope. These two facts account for the now widely accepted observation that friction between macroscopically smooth surfaces increases with the mean absolute surface slope.

## 2.5 COMPARISONS OF THEORIES OF SLIDING FRICTION

### 2.5.1 Energy Dissipation at Asperity Junctions

There are only three basic mechanisms which could possibly cause appreciable energy dissipation at asperity junctions – i.e. fracture, elastic deformation and plastic deformation.

New fracture surfaces will be created whenever an asperity junction is broken, whether at the original interface or within one of the asperities, and whenever a wear particle is formed. However, as will be shown in Chapter 3, except in the case of very severe wear, asperities are normally subjected to many thousands of stress cycles before a wear particle is formed. Therefore, the energy losses associated with wear particle formation will, in general, be negligible, and the contribution of the fracture mechanism to friction will be largely determined by the energy dissipated in breaking the adhered junctions, either at the original interface or within one of the contacting asperities. For brittle materials this will correspond to the energy required to create the two new surfaces, but for ductile metals, which we are considering here, it will be almost entirely determined by the energy consumed in plastic deformation prior to fracture.

Elastic hysteresis and viscoelastic losses can be dominant factors in the friction of elastomers and polymers, and they will be discussed later in this chapter. However, the above discussion of friction theories was restricted to ideal elastic–plastic solids and there can be no energy loss during the elastic deformation of such materials. As many metals are almost perfectly elastic, it is normally assumed that elastic hysteresis can also be ignored as a contributory factor to the observed sliding friction of metals.

Therefore, we must conclude that plastic deformation is the principal cause of metallic friction. At first sight this is a rather surprising conclusion, since, in general, the outermost asperities will always undergo fully plastic deformation during, at most, the first few contacts; they will then have flattened sufficiently to remain nominally elastic. However, on subsequent contact the required elastic stress will be similar to that attained on the first contact – i.e. close to the yield pressure. If we regard the asperities as having spherical geometries, we can see that we now have highly loaded Hertzian contacts, and the surface pressure will be sufficiently high for the maximum shear stress to exceed the critical shear stress for the onset of plasticity. Under normal loading only, the critical shear stress will occur below the surface and the plastic zone will be constrained. This will also be the case if the local friction coefficient – i.e. the ratio between shear stress and normal stress on the asperity – is less than 0.25. In these cases plastic deformation must be very limited, but the energy consumption will still make some contribution to the total friction force. However, if $\mu$ is greater than 0.25, the plastic zone will be at the surface, the deformation will not be constrained and it is likely that a more significant amount of plastic work will be done on each pass of an opposing asperity.

Thus, even when the contact geometry is such that deformation would be expected to be elastic, there may be a significant plastic deformation contribution to the observed frictional work, and, as stated in sub-section 2.4.2, this must be the case when the interfacial shear strength is high, even when the direct stress is very low.

The above arguments will apply to some extent to any asperity which is subjected to an element of plastic deformation (i.e. to a normal pressure greater than $0.6H$), so that a high proportion of the contacting asperities will be subjected to the repeated plastic strain described above.

To assess the relative importance of the various contributions to the total friction, we remember that it takes many thousands of contacts to produce a wear particle. Only very few of the early contacts involve gross plastic deformation and only one – the final one – involves fracture, so that at any time the overwhelming majority of asperity contacts will be nominally elastic, with about half of them experiencing stresses close to the yield pressure. It is inconceivable that the very small fraction of fully plastic contacts will make a significant contribution to the overall friction force, so the deformation contribution to friction will arise almost entirely from localised plastic deformation of the nominally elastic contacts.

The energy dissipation mechanism described above is identical with that described by Johnson for rolling friction under high stresses. We shall refer to it again in Section 2.8, when we discuss rolling friction.

## 2.5.2  The Interaction of Deformation and Adhesion

As we know, when any two surfaces are placed together under a normal load, the surfaces approach each other until the elastic reactive force at the interface is just sufficient to support the normal load. During sliding, a similar situation obtains, with the reactive force being required to support both the normal and the tangential forces.

Force equilibrium demands that each of the asperities forming a junction must be subjected to the same stress, and behaviour during sliding is therefore very dependent on the relative yield pressures of the two materials. If the materials are identical, then they will be strained in the same way, either elastically or plastically, depending on their degree of overlap. However, if their yield pressures are significantly different, the softer material may deform plastically but the deformation of the harder one must be nominally elastic.

As deformation continues, the softer material may work harden; but as long as its current yield pressure is lower than that of the opposing surface, the deformation of the harder material must remain nominally elastic.

It is important to note that the softer material may sometimes cause *localised* plastic deformation in the harder material. This is because the pressure to cause full plasticity of an asperity is approximately twice as great as that required for initiation of subsurface plasticity. Thus, an interfacial pressure which is sufficiently high to cause full plasticity of the softer asperity will also initiate plastic deformation in the harder asperity as long as the hardnesses do not differ by a factor of more than about 2. This will be seen to be particularly important when we come to discuss fatigue wear in Chapter 3.

If one material is very much harder than the other, the condition of stress similarity can only be obtained if the harder asperity penetrates the softer surface, with a consequent ploughing contribution to the friction force. Ploughing is the extreme form of plastic deformation: in what follows we shall consider the less extreme forms of plastic deformation during asperity contact.

For clarity, we shall consider the total life of an asperity junction to comprise three stages: formation, shearing and separation. This is a rather unrealistic picture, because the surfaces are macroscopically in continuous relative motion; therefore, the processes of formation and shear will overlap, as will the processes of shear and separation. Nevertheless, it is a useful way of illustrating the way in which adhesion and deformation interact. The three stages can be characterised by the sense of the direct stress normal to the interface: during junction formation the direct stress is compressive; during shear it progressively falls to zero; and during separation, in the presence of adhesion, it becomes tensile.

During junction formation the only significant process of energy dissipation will be plastic deformation of one or both asperities; the actual amount of energy dissipated will depend on the yield pressures and work-hardening characteristics of the plastically deformed asperities, and on the plastic strains which occur. This plastic deformation will result from the combined effects of the normal and shear stresses, as described in sub-section 2.4.2, and must continue until the total deformation is sufficient to allow the asperities to pass at constant surface separation. The plastic deformation can result in significant junction growth, and may also disrupt any protective surface films, so that strong adhesive bonds are formed over part of the junction.

If the interface is weak, sliding may occur at the interface and the only adhesive contribution to the friction will be the small force required to shear this weak interface. However, if the interface is strong, owing to the formation of strong adhesive bonds, plastic deformation may continue, initially under compression, then in pure shear and finally, as the asperities begin to separate, under an increasing tensile force. As pointed out in sub-section 2.4.2, this plastic shear will occur when the interface is strong, even if the normal stress is well within the elastic range.

It is during junction separation that the surface pull-off force and pull-off work become important and may make an important adhesive contribution to the total friction.

It is interesting to note that interfaces having the same pull-off force may have very different values of pull-off work. One reason for this is illustrated in Figure 2.13, where the stress–strain curves to failure are plotted for a brittle junction and a ductile junction of the same strength – i.e. having the same pull-off force. It can be seen that far more energy will be consumed in breaking the ductile junction and, although both junctions will break at the same force, the ductile junction must make a greater contribution to the observed friction force. This apparent contradiction is resolved when it is realised that the ductile junction will persist over a much longer distance than the brittle one, and therefore its integrated effect on the total observed force will be comparatively great.

It should be clear from the above discussion that the adhesion and deformation theories of friction are not competitive but complementary. There will usually be both adhesion and deformation contributions to any observed friction force. Furthermore, the energy consumed in deformation during junction separation will depend on the strength of the adhesive bond, so the two components are not simply additive, but interactive.

Finally, the products of wear are often abrasive; they may comprise oxide or highly work-hardened metal, or mixtures of both, and they may exist as particles trapped between the surfaces or transferred patches on the rubbing surfaces. In any case, they may cause continuous roughening of one or both surfaces, so that the surfaces never settle down into the nominally elastic condition described above; and if they are very angular, they may give a very significant ploughing contribution to the total friction force.

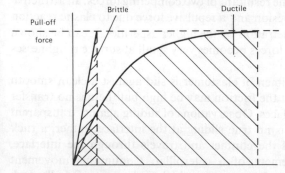

**Figure 2.13** Comparison of pull-off work for brittle and ductile materials having the same pull-off force

The above description of possible forms of behaviour illustrates why it may never be possible to predict accurately the tribological behaviour of a previously untried pair of materials, particularly in the case of unlubricated sliding of metals.

In recent years work has begun on explaining the separation of friction junctions using the ideas of fracture mechanics. This work is very interesting and may well be fruitful, but it is still in its early stages and is beyond the scope of the present chapter.

## 2.6 FRICTION OF NON-METALLIC MATERIALS

### 2.6.1 Elastomers

Elastomers consist of long-chain organic molecules linked at various points along their lengths by strong chemical bonds, the elastic modulus increasing with the number of cross-linking bonds. The free surface of an elastomer consists almost entirely of fully saturated hydrocarbon groups, so that the surface energy is very small. As all the primary interatomic bonds in the surface are fully saturated, the forces between elastomer surfaces are limited to weak van der Waals bonds, rather than the strong cross-linking bonds which exist within the bulk of the material. Thus, in adhesion experiments separation always occurs at the original interface. However, although the thermodynamic work of adhesion, as defined in Equation (2.5), is very low, the pull-off work can be high. This is because there are substantial viscoelastic hysteresis losses as an elastomer is deformed, and the contribution of these losses to the observed work of adhesion is similar to that described for plastic deformation of metals. However, unlike metals, the losses in elastomers are strongly dependent on both temperature and strain rate.

The pull-off force is also critically dependent on the surface roughness of the counterface, and decreases rapidly as the roughness increases. In essence, the pull-off force is the resultant of two competing forces: an attractive force due to the work of adhesion and a repulsive force due to elastic reaction of the elastomer against penetration by the higher asperities. It can be shown that the ratio between these forces becomes very small at surface roughnesses as low as 1 $\mu$m RMS.

If a clean smooth specimen of elastomer is slid against a clean smooth hard surface, it is found that the friction may be high but there is no transfer of material between the surfaces. Observations of sliding against transparent substrates show that there is no true sliding at the interface; rather, a ruck forms towards one edge of the contact and travels through the interface, thus resembling the movement of a caterpillar. Continuous movement proceeds through a sucession of these waves, which are known as Schallamach waves, so that the frictional work is the net effect of the work of adhesion and the pull-off work (Briggs and Briscoe, 1975).

As the degree of cross-linking in an elastomer increases, both the elastic modulus and the mechanical strength increase very rapidly, while the interfacial forces, which are still dominated by weak van der Waals forces, remain approximately constant. Under these conditions the Schallamach waves decrease in size, and it is probable that at some critical point true sliding begins.

Under unlubricated conditions the friction is dominated by the viscoelastic effects associated with the movement of Schallamach waves, the deformation by counterface asperities having little effect. However, in the presence of lubricant the adhesion and the associated viscoelastic losses become very low and the friction is largely dependent on deformation losses. This has led to the use of rubbers of high hysteresis loss in the manufacture of tyre treads, to increase skid resistance on wet roads (Bowden and Tabor, 1964).

## 2.6.2 Polymers

As with elastomers, it has been observed that when a polymer adheres to a hard counterface, the pull-off work is proportional to the thermodynamic work of adhesion, but very much greater. The high values of pull-off work are also associated with deformation losses, but in this case the losses are due to flow of the polymer rather than hysteresis losses.

As with metals, the friction of polymers involves both adhesion and deformation, but in the case of polymers there is little or no junction growth. Also, as with elastomers, the deformation losses during sliding are much more dependent on temperature and strain rate than are those in metals.

Polymers generally have low melting or softening points and poor thermal conductivity, so, when there is significant frictional heating, the surface layer usually softens or melts. The friction then largely arises from viscous losses in the surface layers. At low sliding speeds, where frictional heating is negligible, the frictional characteristics depend on the roughness of the counterface.

When the counterface is smooth, polymers may be divided into two groups. With strongly cross-linked polymers, and glassy polymers below their glass-transition temperatures, the internal bonds within the polymer are usually stronger than the interfacial bonds, so that sliding takes place at the original interface; with softer, semicrystalline polymers, such as PTFE and high-density polyethylene, a thin film or highly oriented polymer is drawn out and transferred to the counterface, and subsequent sliding takes place on this transfer film. In the latter case, if the sliding speed exceeds a low critical value, it becomes easier to rupture the polymer than to draw out a film; the transferred material and loose wear debris then become lumpy and the friction increases.

When the counterface is rough, the asperities can penetrate the polymer and sliding causes the polymer to shear at an angle to the interface. The shear angle correlates with the energy-to-rupture of the polymer, and the frictional work is largely accounted for by the associated energy input.

### 2.6.3  Ceramics

The adhesion of ceramics to each other or to metal counterfaces is not well understood.

The type of primary bonding within the ceramic is clearly important: if it is ionic, then the bonding across the interface will be Coulombic and, in principle, amenable to calculation; however, if the bonding within the ceramic is covalent, it is by no means certain what, if any, valence bonds will be available for interfacial bonding.

Furthermore, as ceramics are elastic solids, they are subjected to the same effect of surface roughness as that described for elastomers in sub-section 2.6.1 – i.e. there is a repulsive force due to elastic reaction of the ceramic against indenting asperities. Thus, the adhesion of ceramics to each other would be expected to be negligible at surface roughnesses as low as a few atomic spacings. However, surface roughness would not have the same effect on the adhesion between a ceramic and a ductile metal, as the effect of asperity penetration would be accommodated by plastic deformation of the metal.

The same problems arise in trying to understand frictional behaviour. If all surface valences are saturated, by either internal bonding or bonding to adsorbed films, it is likely that sliding will take place at the interface and both adhesion and friction will be low. However, it is possible that the sliding process itself could create interfacial bonds which would cause the adhesion and friction to increase.

Despite this lack of fundamental understanding, ceramics and ceramic-based materials are coming into increasing use for tribological applications. It is, therefore, likely that the friction and wear of ceramics will become a major field of research in the near future.

## 2.7  BOUNDARY LUBRICATION

So far in this chapter we have considered sliding between unlubricated surfaces. Later chapters will consider different types of full fluid film lubrication, where sliding behaviour is determined largely by the lubricant viscosity, and elastohydrodynamic lubrication, where sliding behaviour is determined by lubricant properties and the elasticities of the opposing surfaces.

Boundary lubrication is defined in OECD (1968) as 'a condition of lubrication in which the friction and wear between two surfaces in relative motion are determined by the properties of the surfaces, and by the properties of the lubricant other than bulk viscosity'. It is, therefore, a rather ill-defined regime which covers the operating range between dry sliding and elasto-hydrodynamic lubrication: a fluid film is present but it is not sufficiently thick to prevent contact between the higher asperities.

The situation can be characterised by the value of the $\Lambda$ ratio, $h/\sigma$, where $h$ is the separation of the mean planes of the two surfaces and $\sigma$ is the composite roughness $(\sigma_1^2 + \sigma_2^2)^{\frac{1}{2}}$. As we saw in Chapter 1, for surfaces having Gaussian or near-Gaussian asperity height distributions, surface separation is virtually complete if $\Lambda \geqslant 3$. For values of $\Lambda$ less than 3, some asperity contact will occur, and both friction and wear will be expected to increase as $\Lambda$ decreases.

It should be clear from the earlier parts of this chapter that, if we wish to ameliorate the effects of this asperity contact, we need to interpose a film of low shear strength between the sliding surfaces. Therefore, oils intended for boundary lubrication contain additives which will deposit such films on the surfaces.

The most common type of additive is the aliphatic 'fatty' acid of the general formula $C_nH_{2n+1}COOH$, a typical example being stearic acid. Each molecule of such an acid consists of a long covalently bonded hydrocarbon chain with a single reactive acid group at one end. The acid groups of these molecules react with surface metal atoms to form a strongly bonded metallic soap, and the hydrocarbon chains align themselves normal to the surface, thus providing a very effective barrier to metal–metal contact. The process is illustrated schematically in Figure 2.14.

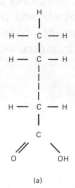

(a)

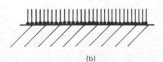

(b)

**Figure 2.14**   Boundary lubrication by long-chain fatty acid molecules: (a) molecular structure; (b) surface protection by reacted layer

These films can provide excellent protection at high pressures and at high sliding velocities, and, as long as the oil does not become depleted of the additive, the soap films should form at any area of exposed metal. The factors which limit the use of these films are not well understood, but it is probable that the most critical factor is the local temperature, which, in turn, depends on both the bulk temperature and the rate of frictional heating.

When the operating conditions become too severe for these boundary lubricants, recourse must be had to extreme pressure (EP) additives. These are aggressively reactive compounds, usually containing sulphur, chlorine or phosphorus, which, again, form protective reaction products at exposed metal surfaces. The components are so aggressive that they cause significant rates of metal removal by corrosion (corrosive wear), and the additive should be selected with care to ensure that its net effect is beneficial.

## 2.8   FRICTION IN ROLLING CONTACT

Lubricated rolling contact will be discussed in detail in Chapter 7. It is the purpose of this section to describe frictional effects during rolling. These effects have been described in detail by Johnson (1985) and only the more salient points will be covered here.

To describe the various effects which can arise, we shall use the coordinate system adopted by Johnson and illustrated in Figure 2.15. This figure shows two non-conforming bodies making contact under a negligibly small force, so that they touch at a single point, $O$. This point is taken as the origin of a rectangular coordinate system $Oxyz$, with the axis $Oz$ coinciding with the common normal to the two surfaces at $O$, and the axes $Ox$ and $Oy$ being chosen, where possible, to coincide with axes of symmetry of the surface profiles. In the special case of line contact between two cylinders, we choose the $y$ axis to be parallel to the axes of the cylinders.

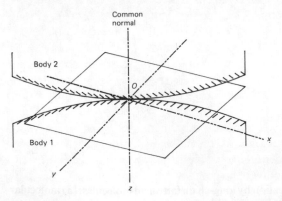

**Figure 2.15**   Non-conforming surfaces in contact at $O$

The frame of reference as defined above moves with the linear velocity of the contact point $O$ and rotates with an angular velocity which maintains its orientation relative to the tangent plane and the common normal at $O$. We can then state the following definitions.

*Sliding* is the relative linear velocity between the two surfaces at $O$; as there can be no component of relative velocity along the normal, sliding is always in the tangent plane.

*Rolling* is the relative angular velocity between the two bodies about an axis lying in the tangent plane.

*Spin* is the relative angular velocity between the two bodies about the common normal through $O$.

Any relative motion between two contacting bodies can be regarded as a combination of sliding, rolling and spin. In the previous sections we have been concerned only with sliding, but in the remainder of this section we shall describe frictional effects due to combinations of sliding, rolling and spin.

To understand such effects, we must also understand the stresses which can arise in the contacting bodies due to the forces acting at the interface. In Chapter 1 we considered in some detail the stresses which arise at contacting non-conforming surfaces due to normal forces acting at the interface, but we referred only in passing to the stresses which arise due to tangential forces. We now need to consider in rather more detail the combined effects of normal and tangential forces.

The resultant force transmitted across any interface can always be resolved into a normal force, $P$, acting along the common normal and a tangential force, $Q$, acting in the tangential plane. The tangential force $Q$ is reacted by frictional resistance and its magnitude must therefore be less than or equal to $\mu P$. When $Q$ attains the limiting value of $\mu P$, gross sliding occurs and the direction of $Q$ is directly opposed to the direction of sliding.

The interfacial forces lead to a contact area of finite size, and the distribution of forces over this area can, therefore, lead to the transmission of a moment across the interface. The components of this moment about axes lying in the tangent plane are known as rolling moments, while the component about the common normal is known as the spin moment.

Following Johnson (1985), we define *free rolling* as rolling motion in which the tangential force $Q$ is zero, and *tractive rolling* as rolling motion in which the tangential force $Q$ is non-zero but less than the limiting value of $\mu P$ at which gross sliding will occur.

The interfacial forces and moments described above are transmitted across the interface by surface tractions, the normal traction, which is simply the contact pressure, being denoted by $p$ and the tangential traction by $q$. In general, both $p$ and $q$ vary over the contact area, so they can be written, respectively, as $p(x, y)$ and $q(x, y)$, where $(x, y)$ denotes position in the tangent plane. It is important to realise that tangential tractions can be present even when the tangential force $Q$ is zero; in such cases it is simply necessary that the tractions integrated over the contact area should sum to zero.

As $p(x, y)$ and $q(x, y)$ have different distributions over the contact area, the ratio $q(x, y)/p(x, y)$ also varies with position within the contact area. As a result, two different contact conditions may obtain: (1) wherever $q(x, y) < \mu p(x, y)$, local sliding is prevented by friction and the corresponding regions of the interface are said to be regions of 'stick'; (2) wherever $q(x, y) = \mu p(x, y)$, local sliding is possible and the corresponding regions are said to be regions of 'slip' or 'microslip'.

A difference between the tangential strains in the two bodies in the 'stick' area leads to a small apparent slip known as creep. Creep can be understood by considering the example of a deformable cylinder rolling under load on a relatively rigid surface. If the tangential strain in the cylinder is tensile, the cylinder is stretched in the 'stick' area. The cylinder then behaves as though it has an enlarged circumference, and therefore in one revolution it moves forward by a distance greater than its actual circumference. The ratio between the distance traversed in one revolution and the actual, unstretched circumference is known as the creep ratio.

The actual location of the boundaries between the areas of 'stick' and 'slip' is quite difficult and need not concern us here. Readers requiring a fuller treatment of such problems are referred to Johnson (1985). In the following sections we shall simply describe in qualitative terms the locations and sizes of the zones of stick and slip for various rolling situations, starting with the most simple situation of free rolling between identical elastic bodies and progressively introducing more variables.

## 2.8.1  Free Rolling between Identical Elastic Bodies

The simplest form of free rolling occurs between two bodies which have the same elastic properties and are geometrically identical. When such bodies roll freely under the action of a normal force, no tangential tractions or slip can occur and the contact stresses and deformations can be calculated from the Hertz equations for normal loading, which are given in Chapter 1.

## 2.8.2  Tractive Rolling of Non-conforming Bodies

We shall largely confine our attention to tractive rolling between bodies having the same elastic properties. However, before considering rolling, it is useful to discuss briefly the tractions which arise at the interface between two stationary cylinders when they are submitted to a tangential force, $Q$, which is less than the force required to cause gross sliding. In this case, within the contact zone, there is a central 'stick' area, where $q(x, y) < \mu p(x, y)$, and two outer 'slip' areas, where $q(x, y) = \mu p(x, y)$. The coexistence of zones of stick and slip is possible because of the deformable nature of the materials.

Thus, if $P$ is kept constant and $Q$ is increased steadily from zero, microslip begins at the edges of the contact zone and spreads inwards as $Q$ increases; when $Q$ reaches the limiting value, $\mu P$, the two zones of slip meet in the centre and gross sliding occurs.

The conditions which obtain in tractive rolling of elastically similar cylinders are qualitatively similar to those described above, with regions of both 'stick' and 'slip' as long as $Q < \mu P$. However, if we examine the position in detail, we find that, in the forward region of slip, the slip would be in the same direction as the tangential traction. This is inadmissible, as it would contravene the basic law that friction must act to oppose the motion. Therefore, in practice, the boundary of the stick region coincides with the leading edge of the contact area and the slip is confined to a single zone at the trailing edge. Again, as $Q$ increases, this slip zone extends forward until, when $Q = \mu P$, the whole area becomes a slip zone and gross sliding occurs.

The problem of rolling between general three-dimensional bodies becomes mathematically very complex. However, the technologically important problem of rolling of spheres has been studied in some detail, and the situation is found to be qualitatively similar to that described for rolling cylinders, with a single stick zone at the leading edge of the contact and slip over the remainder.

### 2.8.3 Sphere Rolling in a Grooved Track

So far we have considered situations where the contact area lies substantially in a single plane. However, in many engineering situations, particularly ball bearings of all types, this situation does not obtain. For a ball rolling along a grooved track, as illustrated in Figure 2.16, the contact area will be an

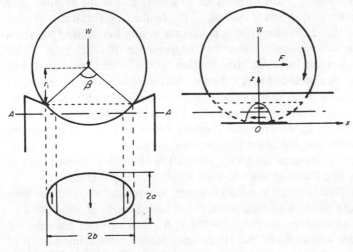

**Figure 2.16** Heathcote-type slip

ellipse and, provided that the angle $\beta$ is less than $30°$, the dimensions of the ellipse are still predicted with reasonable accuracy by the Hertz equations.

If we now consider the situation when the ball has rolled forward through one complete revolution, we can see that in the centre of the groove the ball will have measured out its full diameter along the track, while points on the ball at the edges of the contact zone will have measured out a smaller distance corresponding to the circumference of the circle of radius $r_i$. These differences can be accommodated by the ball slipping in its track, and we might expect the contact area to consist of three zones of slip, as shown in Figure 2.16 – i.e. a single central zone of backward slip and two outer zones of forward slip. In this situation the axis $AA$, which passes through the two boundaries between the regions of forward and reverse slip, can be considered to be the instantaneous axis of rotation. This situation, which is known as Heathcote slip, was first put forward by Heathcote (1921) as it is described above. However, Johnson (1985) has shown that some of the tendency to slip can be accommodated by elastic deformation of the contact areas.

For problems involving both rolling and spin, as in thrust ball bearings, and for problems of materials having dissimilar elastic properties, the solutions become much more complex than those described above. The problems have been treated in detail by Johnson (1985), but are beyond the scope of this book.

### 2.8.4  Rolling Resistance

In all rolling contact situations there is some resistance to rolling, although the magnitude of the resistance is usually very much lower than the resistance to sliding. As with sliding friction, this resistance can arise from only two sources: adhesive losses, due to resistance to sliding in the slip regions, and deformation losses during stress cycling of the contacting surfaces.

It has been found that the rolling resistance in any free-rolling situation is influenced to only a small extent by the presence of lubricants, which would act to lower the adhesive contribution to the rolling resistance, and it can, therefore, be concluded that the contribution of adhesion is very small and that resistance to free-rolling arises mainly from deformation losses. The same is usually true in tractive rolling and rolling in grooved tracks, although, as the tractive force approaches the limiting value or if the spin moment becomes high, adhesive losses become more significant.

In considering deformation losses we shall consider separately losses in elastic contacts and losses in the presence of plastic deformation.

As an example of an elastic rolling contact, we consider a cylinder rolling freely on a plane under a normal load $P$ per unit length, as illustrated in Figure 2.17. The local pressure, $p$, in the contact zone, and the contact semiwidth, $a$, are known from the Hertz equations. Considering a strip of width $dx$ at point $x$, we can see that, during forward compression of the

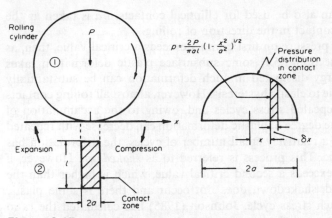

**Figure 2.17**

contacting material, the normal force per unit length of the roller will be $p\,dx$, and this will exert a resistive moment about the centre of the contact zone of magnitude $px\,dx$. Thus, the total moment exerted by all the elements on the front half of the roller will be

$$M = \int_0^a px\,dx = 2Pa/3\pi$$

If the cylinder rolls forward through a distance $x$, the elastic work done by this couple is given by

$$\varphi = Mx/R = 2Pax/3\pi R$$

Under ideal conditions the cylinder would be subjected to an equal and opposite moment arising from the pressure distribution in the rear half of the contact. However, owing to elastic hysteresis losses, not all the work is recovered and there is some dissipation of energy. If we define the hysteresis loss by a coefficient $\varepsilon$, we can see that the energy dissipated in rolling a distance $x$ will be $\varepsilon\varphi$, and if this loss represents the rolling resistance, it follows that

$$Fx = \varepsilon\varphi = \varepsilon\cdot 2Pax/3\pi R$$

where $F$ is the force required to overcome the resistance. By analogy with sliding friction, the ratio $F/P$ can be defined as the coefficient of rolling resistance, $\lambda$, where

$$\lambda = 2\varepsilon a/3\pi R \tag{2.21}$$

Using a method similar to the above, it can be shown that the coefficient of rolling resistance for a sphere rolling on a plane is given by

$$\lambda = 3\varepsilon a/16R \tag{2.22}$$

Equation (2.20) can also be used for elliptical contacts if $a$ is taken as the half-width of the contact in the direction of rolling.

If the contact pressure on first contact exceeds a critical value, then, as described in sub-section 1.5.1, some subsurface plastic deformation takes place and the energy dissipated in such deformation can be substantially higher than that due to elastic hysteresis. However, almost all rolling contacts are subjected to repeated stress cycles and, owing to the accumulation of residual stresses, the degree of plastic deformation can decrease with repeated cycling until, after a relatively small number of cycles, the deformation has become fully elastic. This process is referred to as *shakedown*. However, if the contact stress exceeds a second critical value, which is higher than the value for first yield, shakedown does not occur and there is some plastic deformation on each stress cycle. Johnson (1985) has calculated the ratio between the maximum shakedown pressure $(p_0)_s$ and the pressure for first yield $(p_0)_Y$ for a variety of situations. For free rolling of an elastic cylinder on an elastic–perfectly plastic half-space, this ratio is 1.29, and in the presence of tangential tractions the ratio decreases. For rolling of a sphere on an elastic–plastic half-space, the ratio is 2.2.

When the load exceeds the shakedown limit, some plastic deformation occurs on each cycle and the surface layers are displaced forward relative to the deeper layers. The rolling resistance due to this effect can then be much greater than that due to elastic hysteresis. At even higher loads, when the plastic zone extends to the surface, the rolling resistance increases to even higher values.

## 2.8.5   Tyre–Road Contacts

An important example of rolling wheels subjected to tangential tractions occurs in the wheels of tyres and locomotives. Whether such wheels are being driven or braked, control of the vehicle requires that limiting slip (skidding) does not occur and that the slip behaviour is limited to zones of microslip. For locomotive wheels the previous sections describe the microslip behaviour, but for pneumatic-tyred wheels the situation is more complex and cannot be treated in detail here. However, the principles of the behaviour are given below.

We consider the contact patch between a tyre and a road, as shown in Figure 2.18. We can represent the total friction traction by a single force $Q < \mu P$, acting through some point such as $X$. This force through $X$ can then be resolved into three components acting at the origin $O$:

(1) The force $Q_x$, the driving or braking traction which produces the type of microslip described earlier.
(2) The transverse tangential traction $Q_y$, which will produce transverse microslip; this is the cornering force which produces changes in direction of the vehicle.

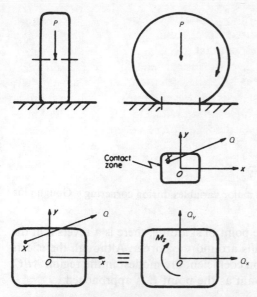

**Figure 2.18**   Resolution of the frictional effects in the tyre–road contact

(3) The torque $M_z$, the self-aligning torque which acts to straighten the course of the vehicle.

When the vehicle is travelling on a straight line, both $Q_y$ and $M_z$ are zero, but if we consider the car being driven round a left-hand turn, the force system and the corresponding microslip velocities are as shown in Figure 2.19. Now the path of motion lags behind the centre line of the tyre by the slip angle $\theta$, which can be interpreted as the angular microslip due to the self-aligning torque $M_z$. The relationships between $Q_y$, $M_z$ and $\theta$ are then shown in Figure 2.20 (Gough, 1954). In the region $ABC$, the self-aligning torque and the slip angle increase with the cornering force. In the region $CD$ there is little change in the self-aligning torque as the cornering force and the slip angle increase, and the feel of the steering is therefore lost. In the

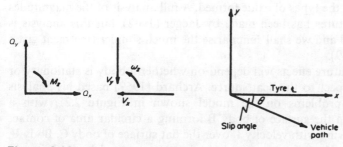

**Figure 2.19**

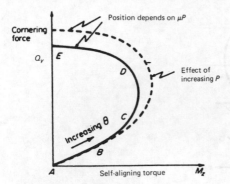

**Figure 2.20**  Relation between the major variables during cornering – Gough plot

region *DE*, which approaches the point of skidding, there is a reversal in the feel of the steering as the car drifts around the corner. Although the effects of the normal load and the friction coefficient are modest in the region *ABC*, they become increasingly important as the point *E* is approached.

## 2.9  THERMAL EFFECTS IN SLIDING FRICTION

Almost all frictional energy is dissipated in the form of heat. As this heat release is a continuous process, temperature gradients will be set up in the contacting bodies, with the highest temperature at the heat source – i.e. the contact surface. The overall temperature rise can, in principle, be calculated from the rate of input of frictional energy and the thermal properties of the contacting bodies and the environment; a theoretical treatment and applications to many practical problems have been discussed by Carslaw and Jaeger (1959). However, as we now know, the bodies make contact only at the asperity tips, and all the heat must be generated at these small areas of contact. As a result, the local temperatures may be instantaneously very much higher than the bulk temperatures of the bodies. These high temperatures of short duration are known as 'flash temperatures', and they can be very important in changing, for example, the microstructures of alloys or the rate of oxidation and the types of oxide formed. A full analysis of the magnitudes of flash temperatures has been made by Jaeger (1942), but this analysis is very complicated and we shall summarise the much simpler treatment given by Archard (1959).

The temperature effects will depend on whether a body is stationary or moving with respect to the heat source. Archard (1959) based his analysis of the various problems on the model shown in Figure 2.21, with a proturberance on the surface of body B forming a circular area of contact $A = \pi a^2$ and moving with velocity $V$ over the flat surface of body C. Body B, therefore, receives heat from a stationary heat source and body C from a

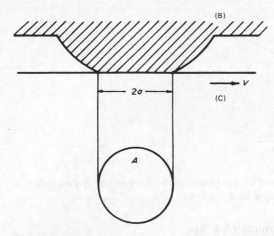

**Figure 2.21**

moving heat source. Then, for the two bodies, $Q_B$ and $Q_C$ are the quantities of heat supplied per second, $K_B$ and $K_C$ are the thermal conductivities, $c_B$ and $c_C$ are the specific heats, $\rho_B$ and $\rho_C$ are the densities, and $\chi_B$ and $\chi_C$ (given by $\chi = K/\rho c$) are the thermal diffusivities. $\theta_{mean}$ is the mean temperature of the contact area, taking the temperature of a point far removed from the contact as zero.

Then Archard derives the following temperature rises for the stationary heat source (body B) and a slow-moving heat source (body C) to be, respectively,

$$\theta_{mean} = Q_B/4aK_B \tag{2.23}$$

and

$$\theta_{mean} = Q_C/4aK_C \tag{2.24}$$

At higher speeds Equation (2.24) does not apply, as there is insufficient time for the temperature distribution of a stationary contact to be established in body C. The speed at which this occurs is determined by the dimensionless parameter

$$L = Va/2\chi \tag{2.25}$$

Equation (2.24) applies to a reasonable approximation for $L < 0.1$. For $L > 5$ the temperature rise is given by

$$\theta_{mean} = (0.31Q/K_C a)\cdot(\chi_C/Va)^{\frac{1}{2}} \tag{2.26}$$

We can define a parameter $N = \pi q/\rho cV$, where $q$ is the rate of heat supply per unit area. Then, for a stationary heat source or a very slow-moving heat source ($L < 0.1$),

$$\theta_{mean} = 0.5NL \tag{2.27}$$

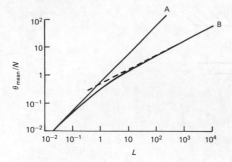

**Figure 2.22** Average temperature as a function of the dimensionless parameter $L$: A, stationary heat source; B, moving heat source

while for a fast-moving heat source $(L > 5)$

$$\theta_{mean} = 0.435NL^{\frac{1}{2}} \tag{2.28}$$

For speeds in the intermediate range $(0.1 < L < 5)$,

$$\theta_{mean} = 0.5\alpha NL \tag{2.29}$$

where $\alpha$ ranges from about 0.85 at $L = 0.1$ to about 0.35 at $L = 5$. These results can be summarised in graphs of $\theta_{mean}/N$ as a function of $L$, as shown in Figure 2.22.

To apply the results shown in this figure, the proportion of the total heat which is supplied to each of the bodies must be taken into account. To do this, the flash temperature is calculated for each body on the assumption that *all* the heat flows into it, using the equation which is appropriate for the speed of the body. Then the true temperature is obtained from the equation

$$1/\theta_{mean} = 1/\theta_B + 1/\theta_C \tag{2.30}$$

Archard applies this theory to show that the maximum flash temperature which can be attained at a steel–steel contact, assuming that the total load is borne by a single plastically deforming area, is of the order of 10 000 °C.

## 2.10 PROBLEMS

(In solving these problems, assume that the friction coefficient is given by $\mu = \mu_{adhesive} + \mu_{ploughing}$.)

1  When a hard steel ball of diameter $D$ is rubbed across a soft metal surface, it ploughs out a groove of width $d$. If $d \ll D$, derive an expression for the ploughing contribution to the friction coefficient.
   $[\mu = d/\pi D]$

2  A hard steel ball is slid across a soft metal surface under two different loads. In one case the friction coefficient is 0.22 and the groove width is

0.3 mm; in the other case the friction coefficient is 0.25 and the groove width is 0.8 mm. Calculate the diameter of the slider and the adhesive contribution to the coefficient of friction.

[$d = 5.3$ mm; $\mu = 0.202$]

3  Two hard conical sliders of semiangles 75° and 80° are slid across an unlubricated metal surface and the measured friction coefficients are in the ratio 11:10. The experiment is then repeated with the surfaces lubricated and the measured friction coefficients are then in the ratio 13:10. Find the coefficients of adhesive friction in the two cases. If the effect of the lubricant is to reduce the surface shear stress by a factor of 3, what are the values of $c$ and $\alpha$ for the unlubricated surface?

[$\mu_{dry} = 0.5938$; $\mu_{lub} = 0.0866$; $c = 0.905$; $\alpha = 13.96$]

# Chapter 3
# Wear

## 3.1 INTRODUCTION

Wear is defined (OECD, 1968) as 'the progressive loss of substance from the operating surface of a body occurring as a result of relative motion at the surface'. This 'loss of substance' is of enormous economic importance. As an example, it was estimated that its cost to the British economy in 1988 would probably be about £2 000 000 000.

Unfortunately, as with friction, there is no reliable way of predicting wear *a priori*, given the two materials to be used and the proposed rubbing conditions. The reader should not find this surprising, as it arises from exactly the same complexity in surface properties and interactions as we have already described in Chapters 1 and 2. However, in the case of wear, the results of making the wrong choice of materials are usually much more serious, owing to the much greater variation in wear rates.

The coefficients of friction observed for most pairs of materials in normal atmospheres are usually in the range 0.1–1, and in the presence of a little lubricant the range is usually narrower than this. Furthermore, if the observed friction coefficient is greater than expected, this usually manifests itself as no more than a small and tolerable increase in power consumption. It is sometimes important that the friction coefficient remain approximately *constant* over a wide operating range (for example, in braking), but this is a separate problem which is always solved empirically.

In contrast, wear rates can vary over many orders of magnitude, and may change catastrophically because of a relatively small change in operating conditions. Thus, the wrong choice of materials can have disastrous consequences.

It is generally agreed that the only safe way to select materials to resist wear in a particular situation is on the basis of tests – first, realistic bench tests, and then field tests. Fortunately, because of the economic importance of wear, many results of such tests have already been published (see, for example, Peterson and Winer, 1980), and potential users can refer to these to make an initial, or sometimes final, choice of materials.

Although there is much current research which is increasing our understanding of wear processes, it is probable that material selection will remain empirical for the foreseeable future. It is true that there are theoretical wear equations, some of which will be described later in this chapter, but they always contain factors which have to be determined by experiment. Furthermore, even these experimentally determined factors are inherently imprecise, and in many cases the user must be satisfied if an observed wear rate is within an order of magnitude of that predicted on the basis of empirical data. The reason for this lack of precision can be described as the dependence of wear rates on extreme values, rather than mean values, of experimental variables. Examples are the domination of the surface chemistry by elements which are only minor impurities in the bulk (Miyoshi and Buckley, 1985); the presence of a few relatively hard inclusions in what is nominally a softer material; the probable disproportionate importance of the heights of the few outermost asperities of a hard material, rather than its average surface roughness; and the presence or absence of contaminants from the environment, particularly in the form of abrasive dust.

However, there are now generally accepted qualitative explanations of the various processes underlying wear behaviour, although there is much that remains to be clarified, particularly the relative importance of the various processes. It would not be possible in a book of this length critically to review the vast amount of work which has taken place in these areas, and in discussing the relative importance of the various processes the views expressed must necessarily be the subjective views of the author. No claim is made for the infallibility of these views, but it is hoped that they do represent a clear and consistent picture of wear, which will provide the interested reader with a background on which to base further reading.

We shall initially concentrate on metals, as they have been subjected to more detailed analysis than have other materials and have a wide range of mechanical properties. In Chapter 4 we shall show how this basic understanding can be extended to the behaviour of other types of materials, particularly ceramics, polymers and elastomers, how it then allows an informed selection of candidate pairs of material to be made and, where possible, how wear behaviour can be predicted on the basis of published data.

## 3.2 TYPES OF WEAR

In a recent review paper Godfrey (1980) states that there are about twelve kinds of wear, each with symptoms which must be recognized for the wear mechanism to be diagnosed. He also states that some authors group several of these types together and that in his opinion this is not justified, as each type has its own mechanism and symptoms. Despite this stricture, for the purpose of clarifying the processes underlying wear behaviour, we shall initially divide these processes into only two types – mechanical and chemical – and shall then follow the conventional approach of subdividing these into fewer than twelve separate processes. Subsequently, we shall attempt to show that this subdivision is somewhat arbitrary, particularly in the case of mechanical wear, and that it is reasonable to think of a continuous spectrum of mechanical wear processes. This is not to belittle the type of approach taken by Godfrey, which may well be a valuable aid to diagnosing and correcting a problem *after wear has taken place*, but we believe that the approach described here will lead to clarity of thought in making a prior selection of materials.

## 3.3   MECHANICAL WEAR PROCESSES

In his excellent 'Survey of possible wear mechanisms', Burwell (1957) listed four major mechanisms: (1) adhesive wear; (2) abrasive wear; (3) surface fatigue; (4) corrosive wear. He also included a fifth classification under the heading 'Minor types of wear', which covered erosion and cavitation. In this section we shall discuss the three mechanisms of mechanical wear listed above, and also the delamination theory of wear, which was put forward some years after the publication of Burwell's review. Corrosive wear is discussed separately in Section 3.6.

### 3.3.1   Adhesive Wear

The theory of adhesive wear has the same basis as the adhesion theory of friction which was described in sub-section 2.3.1. We shall initially follow the historical development of the theory, as it was originally proposed by Bowden and Tabor and formulated as a semi-empirical law by Archard (1953), but in Section 3.4 we shall examine the theory rather more critically and show that some phenomena currently ascribed to an adhesive wear process are explained rather more satisfactorily by a fatigue mechanism.

The basic mechanism underlying the theory has already been explained in sub-section 2.3.1 – i.e. strong cold welds are formed at some asperity junctions and these welds must be sheared for sliding to continue. The amount of wear then depends on where the junction is sheared: if shear takes place at the original interface, then wear is zero, whereas, if shear takes place away from the interface, a fragment of material is transferred from one surface to the other. This transfer of material is observed in practice, normally from the softer material to the harder, but occasionally from the harder to the softer. The theory at this stage has explained only material transfer and not the formation of loose wear debris; in the original theory it was simply stated that subsequent rubbing causes some of the transferred particles to become detached. We shall return to this point in Section 3.4.

### Laws of Adhesive Wear

Here we follow the approach of Archard (1953), who first derived a theoretical expression for the rate of adhesive wear.

It is assumed that the area of contact comprises a number of circular contact spots each of radius $a$. The area of each contact spot is $\pi a^2$ and, assuming, as before, plastic deformation of an ideal elastic–plastic material, each contact supports a load of $\pi a^2 H$, where $H$ is the yield pressure, or hardness. The opposing surface will pass over the asperity in a sliding distance $2a$ and it is assumed that the wear fragment produced from each asperity is hemispherical in shape and of volume $2\pi a^3/3$.

Then the wear volume $\delta Q$ produced by one asperity contact in unit sliding distance is given by

$$\delta Q = (2\pi a^3/3)/2a = \pi a^2/3$$

and the total wear volume, $Q$, per unit sliding distance is $Q = n\pi a^2/3$, where $n$ is the total number of contacts. But each contact supports a load of $\pi a^2 H$, so the total load is $W = n\pi a^2 H$, and $n\pi a^2 = W/H$. Hence,

$$Q = W/3H \qquad (3.1)$$

This equation suggests that there should be three 'laws' of adhesive wear.

(1)  The volume of material worn is proportional to the sliding distance.
(2)  The volume of material worn is proportional to the load.
(3)  The volume of material worn is inversely proportional to the hardness of the softer material.

The first of these 'laws' is found to be true for a wide range of conditions.

The second 'law' is often true from very low loads up to some point at which the rate of wear increases catastrophically. At the point where this transition occurs the apparent pressure (the load divided by the apparent area of contact) is approximately equal to or slightly less than one-third of the hardness. We know from Chapter 1 that $H/3$ is approximately equal to the bulk yield stress. Therefore, this transition clearly corresponds to the point at which the whole surface begins to deform plastically, so that the asperities can no longer be considered to be independent. We also know from Chapter 1 that in the presence of frictional tractions the material will yield at a lower normal stress, and we should, therefore, expect the wear transition to occur at a pressure somewhat lower than $H/3$; this has been confirmed experimentally (Arnell et al., 1975).

The third 'law' is supported by much experimental work, notably that of Kruschov (1957), although, as will be shown later, material properties other than hardness are also important in determining wear rates.

However, there is one important respect in which Equation (3.1) is very misleading. As it stands, it says that the volume of material worn per unit slid distance is *equal* to the normal load divided by three times the hardness, whereas experimental work shows that the worn volume is always less than this, and often by several orders of magnitude. The equation was derived on the assumption that a wear particle was produced at each asperity encounter, and Archard therefore reconciled the theory with experimental observations by postulating that wear particles were produced at only a fraction $k$ of such encounters, so that

$$Q = kW/3H \tag{3.2}$$

or

$$Q = KW/H \tag{3.3}$$

where

$$K = k/3$$

Thus, the uncertainty in predicting a wear rate is due to the uncertainty in the value of $K$, and this must be found experimentally for different combinations of sliding materials and different conditions of rubbing.

The parameter $K$ is known as the wear coefficient, and examination of Equation (3.3) shows that it is dimensionless. However, it can be seen that, even when $K$ is known, it is also necessary to know the hardness before the rate of wear at a particular normal load can be calculated, and the hardnesses of some materials may vary, owing to factors such as work hardening and point-to-point variations in composition. Furthermore, it can be seen that the wear rate depends on $K/H$, rather than $K$ alone, so that tables of $K$ alone must be interpreted with care. For example, a hard material with a high value of $K$ may have a lower wear rate than a softer material with a lower value of $K$. Thus, it is often more convenient to use a wear equation of the form

$$Q = CW \tag{3.4}$$

where the wear constant $C$ can be found experimentally without any independent determination of $H$, and the lowest wear rate will be given by the material with the lowest value of $C$.

### 3.3.2 Abrasive Wear

The term 'abrasive wear' covers two types of situation, known, respectively, as two-body abrasion and three-body abrasion, in each of which a soft surface is ploughed by a relatively hard material. In two-body abrasion a rough

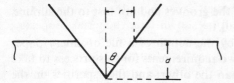

**Figure 3.1**  Abrasive wear by a conical indicator

hard surface slides against a relatively soft opposing surface, whereas in three-body abrasion rough hard particles trapped between the two sliding surfaces cause one or both of them to be abraded.

We initially obtain a semiquantitative expression for the rate of abrasive wear by assuming a simplified model in which a particle of abrasive has a conical shape, defined by the angle $\theta$ shown in Figure 3.1, and the abraded softer surface is flat.

Figure 3.1 shows a single abrasive particle ploughing a track through the softer surface. In traversing unit distance it displaces a volume of material $v = r \cdot d$. But $d = r \cdot \tan \theta$; therefore, $v = r^2 \cdot \tan \theta$. For simplicity, we assume that the softer material has yielded under the normal load only and therefore the abrasive particle transmits a normal load of $\pi r^2 H/2$, where $H$ is the hardness of the softer material.

Thus, if there are $n$ asperity contacts, the total normal load, $W$, is given by $W = n\pi r^2 H/2$, and the total volume of material displaced in unit sliding distance is $Q$, where $Q = nr^2 \cdot \tan \theta$.

Eliminating $n$,

$$Q = 2W \tan \theta / \pi H \tag{3.5}$$

As with adhesive wear, it is found that not all traversals of abrasive particles produce loose wear debris, so Equation (3.5) is modified to

$$Q = k \cdot (\tan \theta / \pi) \cdot W/H \tag{3.6}$$

where $k$ is the proportion of events which actually result in the formation of a wear particle, or

$$Q = K \cdot W/H \tag{3.7}$$

where $K = k \tan \theta / \pi$.

Equation (3.7) is of the same form as the adhesive wear equation (3.1) and, according to this simple derivation, the 'laws' of wear which are derived from Equation (3.1) and given in sub-section 3.3.1 should apply equally well to abrasive wear.

In the derivation described above we assumed that all the material displaced by the abrasive particle becomes loose wear debris, but examination of abraded surfaces shows that this is a gross oversimplification. We shall discuss more modern views of the processes of material removal in Section 3.4. It is sufficient to note here that much of the displaced material can be simply piled up at the sides of the grooves and not lost to the surface. As a result, typical values of $k$ are $\simeq 0.1$.

Two-body abrasion is typified by the action of a file or emery paper on a softer metal surface. There are two requirements for this process to take place: one surface must be harder than the other; and the asperities on the harder surface must be 'sharp' – that is, have high values of $\tan \theta$. The roughness of the surface does not appear in Equation (3.7) but, although the worn volume does not depend on roughness, in practice it is also necessary, if the effects of abrasive wear are not to be too severe, that the hard surface should have a low roughness; this ensures that the abraded surface has a myriad of very fine abrasion marks rather than relatively few very deep and severe ones. Luckily, most finishing processes which reduce asperity slopes also reduce surface roughness.

Severe two-body abrasive wear of lubricated surfaces has been largely eliminated from modern machinery, owing to greater awareness of the importance of material properties and surface finish, and the availability of surface profilometers to ensure that engineering surfaces are of the required quality. Studies of this form of wear are now more related to polishing and grinding processes, where two-body abrasion is deliberately employed. However, in the absence of effective lubrication, when adhesive transfer can take place, both the worn surface and the transferred material can be roughened and work hardened so that two-body abrasion becomes a significant problem.

Three-body abrasion is still an important cause of wear. Many moving parts operate in environments containing significant amounts of adventitious dirt and dust; the products of corrosive wear are more often than not abrasive in character, and metallic wear particles are often fully work hardened, and thus abrasive to their parent surfaces. The practical solution to the problem of three-body abrasion is to flush away abrasive particles and permanently to remove them by filtration.

### 3.3.3  Fatigue Wear

Fatigue wear can be an important phenomenon on two scales: macroscopic and microscopic. Macroscopic fatigue wear occurs at non-conforming loaded surfaces, such as those found in rolling contacts, whereas microscopic fatigue occurs at the contacts between sliding asperities. We shall discuss these two situations in turn.

*Rolling Contact*

Adhesive and abrasive wear depend on direct solid–solid contact and these wear processes operate throughout a period of rubbing. However, if the moving surfaces can be separated by a lubricating film, as in hydrodynamic and hydrostatic bearings, these mechanisms of wear cannot operate. In the absence of any abrasive particles, hydrostatic bearings can be expected to last indefinitely, and hydrodynamic bearings wear only for a fraction of the first revolution after each start-up, so that, for example, the crankshaft bearings in a modern car can be expected to run for $\simeq 10^9$ cycles under severe fluctuating loading without suffering unacceptable wear.

However, in well-lubricated rolling element bearings there is, again, no progressive visible wear due to adhesion or abrasion, but bearing life is limited by fatigue. Such rolling contact fatigue is characterised by the formation of large wear fragments after a critical number of revolutions. Prior to this time there is almost no detectable wear, but as soon as the first wear fragments are formed, the bearing life is at an end. Thus, it is not appropriate to talk of the wear rate of a rolling element bearing; a more relevant term is 'the useful life'.

Although direct solid–solid contact does not occur in a well-lubricated rolling element bearing, the opposing surfaces are subjected to very large stresses, as described in Chapter 2, which are transmitted through the oil film. As we know, in the presence of such stresses the maximum compressive stress occurs at the surface but the maximum shear stress, and therefore the position of first yield, occurs at some distance below the surface. As rolling proceeds, any subsurface element is subjected to a stress cycle for each passage of a ball or roller, and we have already shown in sub-section 2.8.5 that it is this stress cycling which is almost entirely responsible for energy dissipation in rolling contacts. If the stress amplitude is above the fatigue limit for the bearing material, as is almost invariably the case under the high contact pressures used in rolling element bearings, fatigue failure will eventually occur.

The position of failure in a perfect material subjected to pure rolling contact would be defined by the position of maximum shear stress given by the Hertz equations. If some sliding were superimposed on the rolling motion, as in many gear teeth, then the sliding traction would move the position of maximum shear stress closer to the surface. In practice, materials are never perfect and the position of failure would normally depend on microstructural factors such as the presence of inclusions or microcracks. In recent years the lifetimes of rolling element bearings have been improved considerably by using only materials of very high quality to avoid such microstructural effects.

The bearing life, as specified by a manufacturer, is the number of cycles which will be reached or exceeded by 90% of similar bearings under identical operating conditions. It has been found empirically that the life, $L$, is inversely proportional to the cube of the applied load, $W$ – i.e.

$$W^3 \cdot L = \text{constant} \tag{3.8}$$

## Sliding Contact

Wear coefficients in most engineering situations are found to be of the order of $10^{-3}$–$10^{-7}$, with the upper value of $10^{-3}$ normally corresponding to unacceptably high wear. In adhesion theory the wear coefficient is simply defined as the probability of a particular asperity contact resulting in the formation of a wear particle, but this definition leaves unanswered two important questions:

(1) Why, when adhesion theory assumes that all asperities are deformed plastically, are wear coefficients so low?
(2) Why, when all gross plastic deformation of an asperity must usually take place during its first few contacts, do wear particles form after many thousands of further cycles of nominally elastic deformation?

Both these questions can be answered by invoking fatigue as the predominant wear mechanism.

As we now know, when any two solid surfaces are placed in contact under a normal load, the surfaces approach each other until the elastic reactive force at the interface is just sufficient to support the normal load. A similar situation obtains during sliding, with the reactive force being required to support both the normal and the tangential forces. Because of their height distribution the contacting asperities are strained to different extents, so that, in general, the contact stresses will vary from a very low level of elastic stress up to the fracture stress of the weaker of the two contacting materials.

Truly elastic stresses will cause no damage, while the fracture stress will cause the formation of a wear particle during a single interaction. However, at intermediate stresses – i.e. at any stress above the fatigue limit but below the fracture stress – the asperity will be subjected to a single fatigue cycle, and the accumulation of such cycles with continued sliding will eventually cause the formation of a wear particle by local surface fatigue fracture.

(At this point it is useful to discuss briefly the concept of elastic fatigue. To an engineer this is simply taken to mean fatigue of a material at a stress below its yield point. However, to a physical metallurgist true elastic fatigue does not exist: after a cycle of truly elastic deformation, each atom in the specimen would return to its original position, and this stress cycling could be repeated indefinitely without causing fatigue failure. Fatigue failure requires plastic deformation in the form of dislocation movements which eventually lead to the nucleation and growth of fatigue cracks. However, these two views can be reconciled when it is realised that in the polycrystalline materials which are normally tested by engineers some favourably oriented grains can undergo microscopic plastic deformation when the specimen as a whole is subjected to a stress well below the yield stress. The minimum stress at which this localised plastic deformation occurs is known as the fatigue limit – i.e. the stress below which fatigue failure will never occur – and when such a stress exists, it is often only about a half of the bulk yield stress.)

Previous authors have described fatigue theories of wear. Halling (1975) derived a wear equation based on the Manson–Coffin (Tavernelli and Coffin,

1959) fatigue equation, which relates fatigue life to the plastic strain increment during each cycle, and others (Kragelsky, 1965; Soda *et al.*, 1977) have suggested fatigue as a predominant process at low wear rates. In what follows we hope to show that fatigue is probably one of the more important contributory factors in almost all mechanical wear processes.

To do this, we return to the phenomenon of shakedown (Johnson, 1985) which we described in Chapter 2. We recall that, in rolling contact of non-conforming solids, the type of deformation which occurs on repeated loading is critically dependent on the stress level. At stresses below the shakedown stress, contact will become entirely elastic within a few stress cycles, whereas, at any higher stress level, there will be an increment of plastic deformation on each loading cycle.

We should, clearly, expect qualitatively similar behaviour in the sliding contact of non-conforming asperities. There will be a limiting stress below which deformation on repeated cycling will be entirely elastic. However, this stress may be closer to the stress for first yield than is the case in macroscopic rolling contact, because of the difficulty of accumulating in the microscopic contact zones the residual stresses which are necessary for shakedown. At stresses only marginally above the shakedown stress, plastic deformation will be subsurface and constrained, but at higher stresses the zone of plastic deformation will extend to the free surface and be unconstrained. Again, owing to the much higher friction tractions which occur in sliding contacts, the zone of plastic deformation may reach the free surface at much lower stresses than those required in rolling contacts. Indeed, as we showed in Section 2.4, in the presence of strong interfacial bonds plastic flow may occur in shear under very low normal stresses.

In any event, each cycle of plastic deformation will correspond to one fatigue cycle, and the accumulation of such cycles will eventually cause a fatigue fracture and the generation of a wear particle.

Clearly, in assessing the contribution of fatigue to mechanical wear, we need an estimate of the proportion of the contacting asperities which will be subjected to stresses above the fatigue limit, which, for simplicity, we take to be equal to the stress for first yield. This estimate can be made from a more detailed examination of the data summarised in Figure 1.20, which shows the ratio between plastic and elastic contact area, under normal loading only, for different values of the plasticity index and different normalised contact pressures. A selection of the data which were computed to plot these graphs is tabulated in Table 3.1, where $R$ is the ratio between plastic and elastic contact area; $\psi$ is the plasticity index $(E/H) \cdot (\sigma/\beta)^{\frac{1}{2}}$; and $\bar{p}$ is the normalised pressure, $p/H$.

The conclusions to be drawn from the data in Table 3.1 are similar to those of Greenwood and Williamson (1966) – i.e. for all realistic loads the deformation of surfaces is elastic when $\psi$ is less than 0.6, and at least partially plastic when $\psi$ is significantly greater than unity. However, the additional

**Table 3.1**

| $\bar{p} = 10^{-4}$ | | $\bar{p} = 5 \cdot 10^{-4}$ | | $\bar{p} = 10^{-3}$ | | $\bar{p} = 5 \cdot 10^{-3}$ | | $\bar{p} = 10^{-2}$ | |
|---|---|---|---|---|---|---|---|---|---|
| $\psi$ | $R$ | $\psi$ | $R$ | $\psi$ | $R$ | $\psi$ | $R$ | $\psi$ | $R$ |
| 2 | 0 | 2.3 | 5.3 | 5 | 200 | | | | |
| | | 1.3 | 0 | 1.5 | 2 | 5 | 500 | | |
| | | | | 1.2 | 0.5 | 1.3 | 3.6 | 1.8 | 13 |
| | | | | 1.1 | 0 | 1 | 0 | 1.2 | 3 |
| | | | | | | | | 1 | 2 |
| | | | | | | | | 0.6 | 0 |

important point to be noted here is that most engineering surfaces as machined have values of $\psi$ which are substantially greater than unity, and when such surfaces are first placed in contact, the deformation of the contacting asperities will be predominantly plastic. Furthermore, as already noted, the presence of tangential tractions during sliding will exacerbate this situation. It is true that subsequent rubbing may cause the surfaces to run in, so that $\psi$ decreases to a point where the deformation will be nominally elastic but, as long as this nominally elastic deformation is above the fatigue limit, the asperities will be subjected to fatigue cycling. Furthermore, it is clear that surfaces which are machined to a very fine finish, so that the initial value of $\psi$ is less than unity, could still be subjected to similar levels of nominally elastic stress and, therefore, similar levels of subsurface fatigue. Thus, we should expect a high proportion of the contacting asperities on engineering surfaces to be subjected to fatigue wear.

The operation of the fatigue mechanism answers both the questions raised above. First, the wear coefficient can now be simply interpreted in terms of the number of fatigue cycles needed to cause the detachment of a wear particle. For example, if the wear coefficient indicates that only 1 in $10^6$ asperity contacts results in the formation of a wear particle, this suggests that the stress situation is such that, on average, it takes $10^6$ stress cycles to produce a fatigue fracture. Second, although almost all contacts are nominally elastic in the engineering sense, in that no observable irreversible deformation takes place, cyclic, subsurface plastic deformation causes many asperities to fail by fatigue after many cycles of such nominally elastic deformation.

Clearly, it is not necessary for the same asperity pairs to make repeated contact for the above mechanism to operate. In general, the higher asperities will always undergo gross plastic deformation during, at most, the first few contacts, after which they will have flattened sufficiently to remain nominally elastic. However, as before, a high proportion of the contacting asperities will be subjected to stresses which will cause eventual fatigue failure. Furthermore, as the original heavily loaded contacts are worn away, the load is transferred to lower asperities, which, in turn, become subject to fatigue.

In Section 3.4 we shall expand this discussion to include the effects of rubbing dissimilar materials and to take account of interfacial adhesion.

### 3.3.4   Delamination Wear

The delamination theory of wear was first put forward by Suh (1973) and has since been elaborated by Suh (1977) and others (Jahanmir et al., 1974). The theory involves detailed consideration of subsurface dislocation interactions, and readers interested in the details should consult the original papers. For the purposes of this text it is sufficient to set out the physical basis of the theory, which may be summarised as follows (Barwell, 1983).

(1)  When two sliding surfaces interact, the asperities on the softer surface are flattened by repeated plastic loading, until the sliding condition corresponds to that of a harder, rough surface sliding against an approximately plane softer surface. Thus, at each point on the softer surface there is repeated subsurface cyclic shear loading, as already described in discussing fatigue wear.
(2)  As this cyclic loading continues, voids and cracks are nucleated below the surface, crack nucleation at the surface being prevented by the triaxial compressive stress which exists immediately below the surface.
(3)  Further cycling causes these voids and cracks to link up, to form long cracks at an approximately constant distance below the surface.
(4)  When the cracks reach some critical length, the stress situation at the crack tip causes the crack to break through to the free surface, resulting in the formation of the thin plate-like wear particles which are often observed in practice.

## 3.4   DISCUSSION OF MECHANICAL WEAR PROCESSES

In discussing mechanical wear processes we must offer explanations of the full range of wear coefficients, from the maximum possible value of unity to values as low as $10^{-7}$. Although the division is somewhat arbitrary, for clarity we shall initially divide the wear processes into two groups: the higher-wear processes, where the deformation of at least one of the surfaces is seen to be plastic, and the lower-wear processes, where the deformation is nominally elastic.

### 3.4.1   Wear Involving Observable Plastic Deformation

Clearly, the highest wear coefficients are found under conditions of abrasive wear and, under these conditions, the value of $K$ would be unity if every hard asperity removed a chip of material on each contact with the opposing

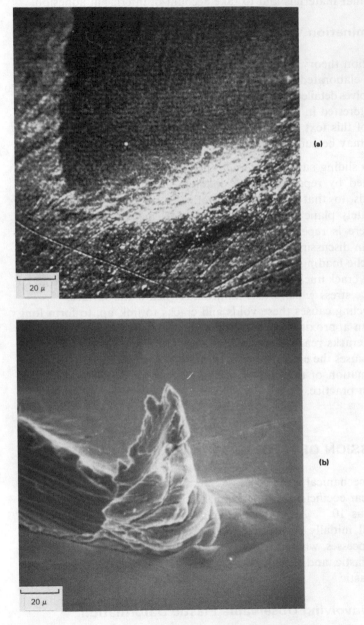

**Figure 3.2**   Effect of asperity shape on the form of abrasive wear (courtesy, Professor
T. H. C. Childs)

(c)

20 μ

**Figure 3.2 (cont.)**

surface. In sub-section 3.3.2 we simply stated the fact that abrasive wear coefficients are normally much lower than this, because much of the displaced material is simply piled up at the sides of the groove and not lost to the surface. In a recent paper Childs (1988) has presented a much more realistic picture of abrasive wear, and this is summarised below.

Studies of metals scratched by model hard asperities generally show that the surface damage is of three types, illustrated in Figure 3.2. The blunter asperities allow the softer metal to flow round them as a wave, as shown in Figure 3.2(a); this does not cause any removal of metal on the first pass, although, as we shall see later, metal may be removed on subsequent passes. The sharper asperities cause chips to be machined from the softer surface, as shown in Figure 3.2(b). Asperities of intermediate shapes cause a wedge or prow of deformed metal to form ahead of them at the start of sliding, as shown in Figure 3.2(c). This wedge effectively blunts the asperity, so that further metal may flow round the wedge without removal (cf. Figure 3.2a), or it may intermittently displace the wedge, to form a new one. Childs mapped the form of material displacement in terms of two variables: the surface slope, $\theta$, and the interfacial shear strength parameter, $\tau_s/k$, where $\tau_s$ is the interfacial shear strength and $k$ is the plastic yield stress in shear of the softer material. (Note that $\tau_s/k$ is identical with the parameter $c$ defined in sub-section 2.3.3, but in this discussion we shall use the form chosen by Childs.)

Childs cites the work of Challen and Oxley (1979), who have developed plane strain slip line field models of wave-forming and cutting flows in friction and wear. They showed that, for a wedge-shaped asperity of slope $\theta$, wave flow should occur when

$$\theta \leqslant 0.5 \cos^{-1}(\tau_s/k)$$

while chip formation should occur when

$$\theta \geqslant \pi/2 - 0.5 \cos^{-1}(\tau_s/k)$$

The limiting conditions represented by the equalities in these equations are mapped as the lines $AB$ and $BC$, respectively, in Figure 3.3. Within the region $ABC$ no steady flow is possible and this is the region of initial wedge formation as illustrated in Figure 3.2(c).

Further work by Challen and Oxley (1984) has shown that the situation is far more complicated than that illustrated in Figure 3.3. However, discussion of this work would not be appropriate here, and readers requiring more information are referred to the original papers.

Childs expresses the severity of wear by a wear coefficient $K$ in an equation similar to the Archard equation for adhesive wear – i.e.

$$V = KWs/H \tag{3.9}$$

where $V$ is wear volume, $W$ is the normal load, $s$ is the slid distance and $H$ is the hardness of the softer material. However, in this case the $K$ value cannot be interpreted as the probability of forming a wear particle, which would have a limiting value of unity: it is simply a coefficient relating the volume of material removed per unit slid distance to the load and the hardness, and can have a value greater than unity.

When metal is removed as a chip, the wear process is described as

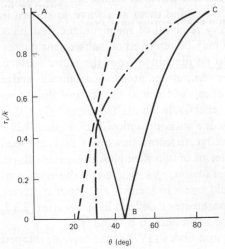

**Figure 3.3** Map of flow regimes (after Childs, 1988)

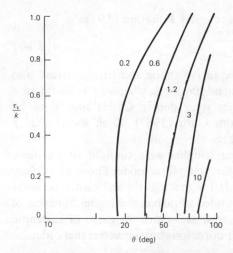

**Figure 3.4**  Calculated values of abrasive wear coefficient $K$ for various values of $\tau s/k$ and $\theta$ (after Childs, 1988)

machining wear. In this regime upper limits to wear coefficients can be calculated, using slip line field theory, and such wear coefficients, for various values of $\tau_s/k$ and $\theta$, are shown in Figure 3.4. It can be seen that the calculated values of $K$ for the machining process fall to zero at a surface slope of approximately 20°, which is very rough by normal engineering standards. Other wear processes must, therefore, take over at lower slopes, and Challen and Oxley have shown that in the region of plastic wave flow the observed wear rates can be explained on the basis of high strain fatigue.

Childs points out that, in the fully plastic conditions to which his work applies, calculations of the forces associated with the flow fields automatically combine the effects of adhesion and deformation, and may therefore be used to calculate friction coefficients. Such calculations have been made, and they show that for wave flows friction coefficients are less than 0.4, while with cutting wear they are greater than 0.4.

The theoretical work described above was based on plane strain deformation by wedge-shaped asperities. However, real engineering situations involve three-dimensional asperities having a wide variety of shapes and orientations, and Childs cites many sources to show that the rate of abrasive wear and the transitions from wave flow to prow formation and from prow formation to cutting are very dependent on the shape and orientation of the abrasive particle and on the friction stress.

The transitions are also strongly dependent on the mechanical properties of the abraded surface, the critical parameters being its hardness and work-hardening characteristics. In addition to altering the flow regime boundaries, work hardening also affects the wear rate within the cutting regime.

To illustrate these effects, Childs rewrites Equation (3.9) as

$$V = fKWs/H \tag{3.10}$$

where $K$ is primarily dependent on abrasive shape and friction stress, and relates the scratch volume to the load and hardness, whereas $f$ is the fraction of the scratch volume which becomes wear debris. Childs cites work by Hokkirigawa and Li (1987) and Zum Gahr (1987) which shows that $f$ decreases as surface hardening increases.

For hard brittle metals fracture toughness is thought to be more important than work hardening in influencing flow mode. The work of Zum Gahr shows that the ranking of $fk/H$ for martensitic and austenitic irons changes with abrasive slope. With rounded abrasives the higher hardness of martensitic irons gives them a better wear resistance than that of austenitic irons, but with sharper abrasives their poorer toughness reverses that order.

### 3.4.2 Wear under Nominally Elastic Contact Conditions

In sub-section 3.3.2 we have already shown that, during nominally elastic contact, a high proportion of the contacting asperities will be subjected to repeated subsurface shear stresses in the plastic range, and, consequently, to eventual fatigue failure. Thus, during nominally elastic contact, which obtains on all well-run-in surfaces, fatigue will be the predominant wear mechanism.

As shown in sub-section 3.3.3, when the plasticity index is less than 0.6, all asperities deform nominally elastically, and in many cases, particularly at low nominal pressures, the deformation may be truly elastic. In such cases wear rates are extremely low and are often due to corrosive wear only or to the removal of material at the atomic level rather than as particles.

Recently Halling (1988) has developed a criterion for elastic contact which has already been described in sub-section 2.3.1, and implicitly used in Table 3.1. The criterion gives the limiting normalised pressure for elastic behaviour as

$$\bar{p} = p/H = \tfrac{4}{3}(\eta\beta\sigma\psi) \int_{(b-(1/\psi^2))}^{b} (s-h)^{\frac{3}{2}}\varphi(s)\,\mathrm{d}s \tag{3.11}$$

where $p$ is the nominal applied pressure; $b$ defines the height, $b\sigma$, of the highest asperity; $s$ is the non-dimensional height, $z/\sigma$, of an asperity; $h$ is the non-dimensional separation, $d/\sigma$, of the surfaces; and $\psi$ is the plasticity index. This criterion was developed for surface contact under normal load only, but it would also apply to frictional loading with friction coefficients less than 0.1. The criterion would need to be modified for situations of higher friction; nevertheless, it still provides a useful insight into the factors which must be considered in choosing materials to resist wear. In particular, it stresses the great importance of the plasticity index.

### 3.4.3   The Form of the Wear Laws

We know from experiment that wear is proportional to the normal load and slid distance, and, although we have shown that this is predicted by theories of adhesion and abrasion, we have not shown that it would be expected if fatigue were the predominant wear process. Fatigue wear equations predicting such relationships have been derived by Halling (1975) and Kragelsky (1965). However, the following simple argument shows that the observed linear relationships of wear rate to both normal load and slid distance derive simply from surface topography and would be expected for *any* form of mechanical wear involving asperity contact only.

(1) As shown in Chapter 1, when contacting surfaces have exponential height distributions, the mean contact pressure and the mean individual contact area are independent of the load: the number of contacts, and thus the total contact area, are proportional to the load. The same is true, to a very good approximation, for surfaces having Gaussian or near-Gaussian height distributions.
(2) Most surfaces have Gaussian or near-Gaussian height distributions, at least in the highest 10% of the surface to which contact is normally limited, so that surfaces under load are subjected to contact stresses and contact stress distributions which are almost independent of the load.
(3) Whatever the mechanical wear process, the probability that an individual asperity interaction will lead to the formation of a wear particle will depend on the local contact stress, and is therefore also independent of the load.
(4) The effect of an increase in the load is simply to give a proportionate increase in the true contact area and, thus, in the number of contacts in each stress state, including the number of wearing contacts.
(5) Similarly, the effect of an increase in slid distance is simply to give a proportionate increase in the number of wearing contacts.

Statements (4) and (5) can be combined to write down a general mechanical wear law as

$$Q = K \cdot A \tag{3.12}$$

or

$$Q = K \cdot W / \beta \tag{3.13}$$

where $\beta$ is a mechanical constant defining the load capacity. For plastic asperity contact this constant is simply $H$, in line with Equation (3.3), whereas with elastic contact its value will be somewhat different from $H$. However, we now know that even in elastic contact most asperities are very close to their yield stresses, so that $\beta$ may be taken as being approximately equal to $H$ for all practical situations.

### 3.4.4 Wear of Dissimilar Materials

When two materials having different hardnesses are loaded together, the two opposed asperities involved in any particular contact must be loaded, and therefore stressed, to the same level. If the hardnesses are significantly different, the softer material may deform plastically, but deformation of the harder material must remain at least nominally elastic. As the load is increased, the plastically deforming material may work harden; but as long as its current hardness is lower than that of the opposing surface, the harder asperity will remain nominally elastic. However, we now know that subsurface plastic deformation will occur when the pressure at the surface is about half the hardness. Therefore, a softer material can cause subsurface plastic deformation in a harder one as long as its hardness, in the fully work-hardened condition, is more than about half the initial hardness of the harder surface. To emphasise this important point, it can be seen that the harder surface must be at least twice as hard as the softer if it is to remain fully elastic.

Thus, behaviour during sliding is very dependent on the difference in hardness between the two materials.

If one material is very much harder than the other, the condition of stress equality can only be obtained if the hard asperity penetrates the softer surface, as in a hardness test. Relative sliding then requires the hard asperity to plough a groove in the opposing softer surface. Hence, the softer surface is subjected to abrasive wear. However, it must be remembered that if the hardness ratio is less than about 2, the abrasive surface may still be subjected to subsurface plastic deformation and fatigue wear.

If the two hardnesses are such that both surfaces can remain nominally elastic (a condition that often obtains after running in), it is probable that each asperity of any contacting pair will be subjected to subsurface plasticity and, hence, fatigue. Indeed, both will be subjected to identical cycles of fatigue stress, and repeated contacts will cause fatigue failures at both surfaces. However, the harder asperity will require more stress cycles to produce a wear particle, and will therefore wear at a lower rate than the softer one. This is illustrated schematically in Figure 3.5, which shows the $S$–log $N$ fatigue curves (stress–log number of cycles to failure) for the two materials. For the example given, a horizontal line through the stress $S_0$ shows that the numbers of cycles to failure for the harder and softer materials at the stress $S_0$ would be in the ratio $N_H/N_S$ and the relative wear rates would thus be in the ratio $N_S/N_H$ (remembering that $N$ is shown on a log scale). This oversimplifies the true situation, where the stress cycles would be of varying amplitude, but it clearly illustrates the principle involved.

Thus, fatigue wear explains the observation that a harder material can be worn by a softer one, a phenomenon which has never been explained satisfactorily by adhesion theory.

By invoking fatigue as the basic wear process we can also explain the formation of loose wear debris.

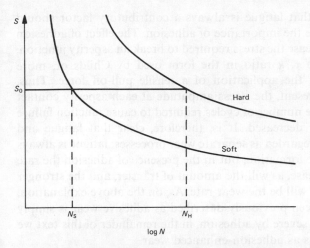

**Figure 3.5** Illustrating simultaneous fatigue of both hard and soft rubbing surfaces

It will be remembered that the adhesion theory explains the transfer of particles of the softer material to the harder surface by postulating that the interfacial bond is so strong that the junction breaks in the softer material rather than at the interface. The theory then goes on to suggest that loose debris is formed by the transferred particle becoming detached during a subsequent contact. However, it is difficult to see how the transferred particle can be detached in this way. For the initial transfer to occur, the bond of the softer material to the harder must have been stronger than the cohesive strength of the softer material in its fully work-hardened condition. Therefore, the softer material should never be sufficiently strong to break the bond between the transferred particle and the counterface. However, if the transferred particle is removed from the softer material by fatigue cycling, its bond to the counterface may be very weak, corresponding only to the force required to cause final fracture during the last fatigue cycle. Indeed, there is no requirement for the observed loose debris ever to have been attached to the counterface: it may have suffered final detachment simply because of a collision with an opposing asperity.

In summary, with the exception of the cutting wear process described by Childs, it is evident that fatigue is a contributory factor over the full spectrum of wear behaviour. This is so by definition, as fatigue failure is defined as failure caused by repeated applications of stresses which are insufficiently high to cause failure in a single application – a situation which manifestly obtains in all wear situations where the wear coefficient is less than unity. Thus, the very high wear rates observed in the milder forms of abrasive wear – i.e. wave-forming and prow-forming – can be ascribed to high strain fatigue, in line with Challen and Oxley, while the low wear rates observed in nominally elastic contact can be ascribed to low strain fatigue.

However, the fact that fatigue is always a contributory factor should not be taken to minimise the importance of adhesion. The effect of adhesion will invariably be to increase the stress required to break an asperity junction, either by increasing the $\tau_s/k$ ratio in the form used by Childs or, more obviously, by requiring the application of a tensile pull-off force. Thus, whenever adhesion is present, the stress amplitude at each asperity contact will be increased, and the number of cycles required to cause junction failure will be correspondingly decreased. It is, therefore, clear that fatigue and adhesion should not be regarded as separate wear processes: fatigue is always the basis of wear particle formation, but in the presence of adhesion the rate of fatigue wear will increase, as will the amount of transfer, and the stronger the adhesion the greater will be the wear rate. As, on the above explanation, the process which has been previously described as adhesive wear is simply fatigue wear made more severe by adhesion, in the remainder of this text we shall refer to this process as adhesion-enhanced wear.

Delamination wear also conforms to the definition of fatigue given above – i.e. failure resulting from many stress cycles at a stress level which cannot cause failure in a single cycle. The distinctive feature of delamination theory is that it invokes fatigue cycling over relatively large areas at an approximately constant distance below the surface, and thus explains the occurrence of the plate-like wear particles which are commonly observed. It is very likely that the fatigue process described by delamination theory does indeed take place. However, an alternative explanation of the plate-like wear particles is that they are formed by severe repeated compression of initially equiaxed wear particles. There is no doubt that when such an equiaxed particle was first formed, it would instantaneously separate the sliding surfaces and thus take a large fraction of the total applied load. The resulting stress would certainly cause very severe plastic compression of the particle in a plane parallel to the interface and would thus deform the initially equiaxed particle into a plate-like shape (Sasada *et al.*, 1979).

## 3.5  CORROSIVE WEAR

When rubbing takes place in a corrosive environment, either liquid or gaseous, then surface reactions can take place and reaction products can be formed on one or both surfaces. These reaction products can be removed by rubbing, and the cycle of reaction and removal can be repeated indefinitely.

By far the most common corrosive medium is the oxygen content of air. Therefore, we shall largely confine our attention to corrosive wear in an oxidising environment, otherwise known as oxidative wear, but the principles described would apply to wear in any other corrosive medium.

The first important point to note is that oxidation of sliding surfaces is usually beneficial. The oxide film prevents metal–metal contact and thus mitigates against the severe adhesion-enhanced wear which would otherwise occur. Very often, when surfaces are oxidised, the wear debris is finely divided oxide, the rubbing surfaces remain smooth and the rate of loss of material is low.

The effects of oxidation depend not only on the oxidation rate, but also on the mechanical properties of both the oxide and the substrate, and the adhesion of the oxide to the substrate. As with the mechanical wear processes discussed earlier, it is unlikely that there will ever be a theory of oxidative wear which will allow accurate predictions of wear rates over a wide range of conditions. Any such theory would require, as before, precise knowledge of the conditions necessary to produce a wear particle. In addition, it would require knowledge of the surface temperature distribution, as a function of both position and time; of the relationship between this distribution and the oxide film thickness; and of the stabilities of films of different thicknesses under the variable thermal and mechanical conditions to which they would be subjected. In particular, as oxidation is a thermally activated process, the rate of oxidation can increase exponentially with temperature. This makes prediction particularly difficult, as a change of only a few degrees in surface temperature can change the rate of oxidation by an order of magnitude.

Simplified models of oxidative wear have been produced (Quinn, 1971; Earles and Powell, 1971) and they probably provide good qualitative descriptions of the wear process. However, to give good agreement with observed wear rates, it is necessary to invoke, for example, activation energies for oxidation which are different from those found in static oxidation experiments. We can conclude that the prediction of oxidative wear rates will remain empirical, and we shall restrict ourselves to a qualitative description of the process.

The oxides of most metals, including iron, have volumes which are significantly different from those of their parent metals. Therefore, during static oxidation, oxide formation causes the development of stresses in the oxide which increase with film thickness. At some critical film thickness the oxide film can then fail, either by blistering, when the stress within the film exceeds the strength of the adhesive bond between oxide and substrate, or by cracking, when the oxide fails in tension. The critical film thickness thus depends on the metal/oxide volume ratio, on the adhesive bond between the oxide and substrate and on the strength of the oxide.

The complexity of the process increases substantially when sliding occurs. In this case failure does not depend only on the stresses resulting from the film formation, but also on those caused by the normal and tangential external loads, including the thermal stresses induced by flash temperatures at asperity contacts.

Finally, the stability of the film will be dependent on the hardness of the substrate: a hard substrate will provide good support for the film, whereas a softer substrate is likely to deform plastically, at least locally, and thus effectively leave the thin brittle film unsupported.

Archard (1980) has given a simple qualitative development of corrosive wear theory which shows that the rate of wear can again be described by an equation of the form $Q = kW/H$, where the wear coefficient, $K$, is given by

$$K = K_3 \lambda / 2a \qquad (3.14)$$

where $K_3$ is the proportion of events which produce a wear particle, $\lambda$ is the critical film thickness at which the film becomes unstable and $2a$ is the width of an asperity contact. The values of both $K_3$ and $\lambda$ are dependent on the various thermal, chemical and mechanical variables described above.

As factors such as reactivity and film thickness are functions of both time and temperature, the effectiveness of oxidation in providing a protective film can vary dramatically for relatively small changes in rubbing conditions. The best-known example of this is in the mild-to-severe wear transitions observed in the unlubricated sliding of many metal pairs. This effect is illustrated schematically in Figure 3.6, which shows a log–log plot of the variation of wear rate with load for carbon steel sliding on carbon steel.

At low loads, below the first transition load, $T_1$, the wear rate remains low, the surfaces remain smooth and the wear debris is finely divided and almost entirely oxidised. It is thought that in this regime the wear debris is removed in the form of oxide, and that the worn areas have time to reoxidise before they next make contact. (Remember that the number of contacts in any stress state is proportional to load, so that the time available for oxidation between contacts is inversely proportional to load.)

At the transition load $T_1$, the wear rate changes by more than two orders of magnitude, the wear debris consists of relatively large metal particles and the surfaces are severely damaged and roughened. In the range $T_1 - T_2$ there is insufficient time for the contact spots to reoxidise between contacts, so that adhesion-enhanced wear takes place.

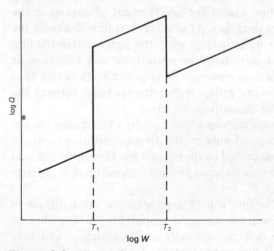

**Figure 3.6** Illustrating the mild-to-severe wear transitions of an unlubricated steel

The rate of oxidation and, hence, the film thickness attained between successive asperity collisions are very dependent on the local temperature, and this temperature, in turn, depends on the input of frictional work and thus on the load. Therefore, all other things being equal, the rate of oxidation increases approximately exponentially with the load, while the time between collisions decreases only linearly. Thus, at the second transition load, $T_2$, the temperature effect becomes predominant and the wear process returns to one of 'mild' oxidative wear.

Steel surfaces examined after rubbing above $T_2$ also have a hardened surface layer, due to the repetitive rapid cooling from high flash temperatures. This hardened layer also favours the mild wear regime, owing to the reduced incidence of plastic deformation and the improved support given to the oxide film. This effect of hardness is illustrated more dramatically by rubbing steel–steel pairs which have been previously hardened and tempered to different bulk hardnesses. As the initial hardness is increased, it is observed that the severe wear regime becomes narrower until, at some critical value of hardness, it disappears entirely and mild wear persists over the full load range.

In the above discussion we have considered only the effect of increasing the load. However, it is clear that similar effects would be seen as a result of increasing the speed. Increasing speed would both increase flash temperatures, owing to the increased input of frictional work, and decrease the time available for reoxidation between collisions.

It has been pointed out by Endo and Fukuda (1965) that the presence of oxygen at crack tips will tend to increase the rate of fatigue crack growth and, thus, the fatigue wear. This is a further example of the interactions between, and complementary effects of, wear processes which are often described as though they are quite independent.

Finally, Barwell (1983) has pointed out that the adjectives 'mild' and 'severe' can be very misleading, in that the rate of material loss in a 'mild' wear regime can be substantially higher than that in a 'severe' wear regime. The terms 'mild' and 'severe' apply to the form of surface damage rather than the rate of material loss, and Barwell suggests that the term 'smooth sliding wear' should be substituted for 'mild wear'.

As mentioned earlier, the above discussion also applies to corrosive media other than oxygen. For example, we have already described, in Section 2.7, the use of extreme-pressure additives to deliberately generate significant corrosive wear in preference to the much more damaging severe wear which would occur in their absence.

## 3.6 WEAR MAPS

Recently Lim and Ashby (1987) have suggested that the various regimes of mechanical and corrosive wear for any particular pair of rubbing materials

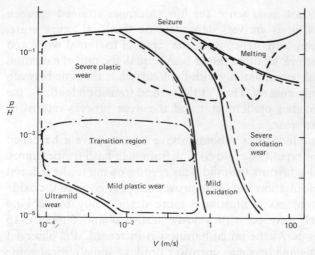

**Figure 3.7** Wear map for soft carbon steels (after Lim and Ashby, 1987, and Childs, 1988)

could be shown on a single wear map plotted on axes of normalised pressure $p/H$ and sliding velocity. As an example of this technique, a wear map for soft carbon steels sliding in air at room temperature is shown in Figure 3.7. It can be seen that, in principle, the map can be divided into areas corresponding to different wear regimes, with boundaries of sliding speed and contact pressure beyond which a purely mechanical view of wear would be grossly in error, owing to the effects of oxidation. Childs (1988) has criticised this approach, at its current stage of development, on two grounds: that it takes no account of differences of surface slope or surface shear strength; and that it is only of the same practical use as quoting a wear coefficient. Nevertheless, although the plotting of such maps for a wide range of environmental conditions would require the analysis of enormous amounts of wear data, they do offer the possibility of condensing such data into a form which would be readily understood by designers, and, together with the alternative type of map proposed by Childs (Figure 3.3), may offer the first real opportunity of developing simple design guides for sliding wear behaviour of metals.

## 3.7 MINOR TYPES OF WEAR

### 3.7.1 Fretting

The wear process known as fretting is not a separate process, but it is convenient to treat it separately. It has been described in detail by Waterhouse (1988), and only the principles of fretting behaviour are summarised here.

Fretting can occur whenever low-amplitude vibratory sliding takes place between two surfaces. This is a common occurrence, since most machinery is subjected to vibration both in transit and in operation.

Fretting can combine many of the wear processes already described. The oscillatory sliding causes fatigue wear, which may be enhanced by adhesion. Most commonly, this wear is combined with the effects of corrosion, by oxygen or some other medium, and, as many corrosion products are harder than their parent metals, this can also lead to abrasion. The fact that there is no macroscopic sliding at fretting contacts often means that the fretting wear debris cannot escape but is trapped between the surfaces.

In fretting there may be no true macroscopic sliding: the surfaces may be in static contact over the central region of a contact zone, where the contact pressure is relatively high, and subject to microslip at the outer regions of the zone, where the pressure is relatively low and the tangential tractions are sufficiently high to overcome the static friction. Under such circumstances, fretting wear takes place over a narrow region at the edge of the contact zone; in this region fatigue cracks may be nucleated and propagated by the oscillatory stress, and this fatigue can lead to gross component failure.

### 3.7.2 Erosion

Erosion is the form of damage experienced by a solid body when liquid or solid particles impinge on a solid surface. In describing erosion, it is useful to describe separately the effects of impacting liquids and solids, although, in practice, the two processes often occur simultaneously.

*Fluid Erosion*

There are two basic types of fluid erosion – i.e. liquid impact erosion and cavitation erosion. The current state of understanding of these processes has recently been comprehensively reviewed in excellent papers by Hammitt (1980) and Schmitt (1980), and it is the purpose of this sub-section to give only an outline of the more salient features of the processes.

The processes of removal of material are rather less well understood than are the processes occurring in sliding wear, but many studies have been made of the damage which occurs. It appears that there are different damage processes for brittle solids, ductile metals and thermoplastic polymers, thermosetting polymers and elastomers.

Liquid impact on a brittle material generates momentary stresses which are sufficiently high to cause cracking in the form of initially unconnected annular ring segments. With continued impact, further cracking occurs and the cracks eventually join, so that material is removed in the form of chips.

The erosion damage of ductile materials, such as many metals and thermoplastic polymers, initially takes the form of surface depressions with upraised edges. These edges are then subjected to high stresses during the radial flow of subsequent impacting drops, until they eventually fail, with consequent surface roughening. The surface roughening itself then changes the flow patterns of impacting liquid, with a possible exacerbation of the erosion.

The erosion behaviour of thermosetting polymers is similar to that already described for brittle materials, with the coalescence of ring cracks leading to the formation of relatively large wear particles.

Elastomers are often used in the form of surface coatings to reduce erosion damage. Some of these coatings fail by erosive wear which is phenomenologically similar to that described for thermosetting plastics, while others remain superficially almost undamaged but suffer loss of adhesion to their substrates.

Cavitation damage occurs when bubbles entrained in a liquid become unstable and implode against the surface of a solid. This is thought to produce a combination of shock waves in the liquid and liquid microjet impact on the solid surface. Cavitation damage closely resembles damage by liquid impact erosion, but it is on a finer scale. It can be a very important and damaging form of attack on solids which are moving in a liquid, examples being ships' propellers and pump impellers.

### Solid Erosion

When a stream of solid particles impinges on a solid surface, it is found that the rate of erosive wear is dependent on the angle of incidence of the particles, and that the wear rates for ductile and brittle materials follow different curves, as shown in Figure 3.8.

For ductile materials the maximum erosion occurs at an angle of approximately 20° to the surface. Several workers (Bitter, 1963; Finnie *et al.*, 1979) have attempted to model this process but there is, as yet, no accepted explanation of the form of the curve. Perhaps the closest approach to an accepted analysis is that due to Bitter. He suggests that there are two wear processes: cutting wear, which predominates at low angles, and deformation wear, similar to that described for liquid erosion, which predominates at high angles. The observed peak at 20° is then the point at which the sum of the two effects has its maximum value.

For brittle materials the erosion process seems to be similar to that described for liquid impact, with wear particles forming due to the coalescence of partial ring cracks.

There are various theories of erosion damage but there are no reliable methods of predicting erosive wear rates. Erosion damage in brittle materials has been modelled as Hertzian impact between elastic solids (Hutchings, 1979) and some degree of agreement between theory and experiment has been obtained. With ductile materials it is well established that there is an

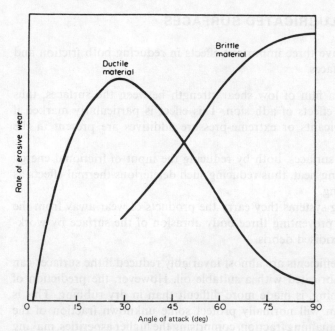

**Figure 3.8**   Dependence of rate of erosion on angle of attack of impinging particles

'incubation' period during which no loose debris is formed, and this suggests that erosive damage of ductile materials is a cumulative process, with fatigue again being a likely cause.

## 3.8   EFFECTS OF SURFACE FILMS ON WEAR

It should now be clear that the wear resistance of any surface is dependent on both the hardness and the elastic modulus of the contacting surface. We have already shown in Section 1.5 that a relatively thin surface film can modify both the hardness and the modulus of a surface, and thus have a very significant effect on its tribological properties. This has recently been treated in detail by Halling (1986), who shows that the critical factors in determining the tribological properties of any substrate–coating–counterface combination are the mechanical properties of the three materials, the roughnesses of the substrate and counterface, and the thickness of the surface film. As both film thickness and surface roughness are very likely to change with continued rubbing, there is unlikely to be a quantitative method of predicting the long-term wear behaviour of coated surfaces in the near future. However, the principles established by Halling are very useful and can guide the selection of coated materials for tribological applications. We shall return to this topic in Chapter 4.

## 3.9   WEAR OF LUBRICATED SURFACES

Fluid lubricants have three important effects in reducing both friction and wear of sliding surfaces.

(1) They provide a film of low shear strength between the surfaces, thus mitigating the effects of adhesion. This effect is particularly marked if boundary lubricants or extreme-pressure additives are present in the lubricant.
(2) They cool the surfaces, both by reducing the input of frictional energy and by removing heat, thus reducing such deleterious thermal effects as surface softening.
(3) In recirculating systems they carry the products of wear away from the interface, thus preventing three-body abrasion of the surface by work-hardened or oxidised debris.

Thus, wear coefficients are almost invariably reduced if the surfaces can be continuously lubricated with a suitable oil. However, the prediction of wear rates, if anything, is made more difficult than in dry rubbing. This is because the oil film will normally protect some unknown fraction of the surfaces, with the remaining fraction, comprising the higher asperities, making solid contact through the film. Thus, the effective wear coefficient is the resultant of one coefficient on the unprotected fraction of the surface, and of a second coefficient on the protected fraction.

As a result of this additional complexity, hardly any wear coefficients have been published for conditions of lubricated sliding. Recently Rowe (1980) has suggested that wear coefficients can remain approximately constant under some sliding conditions, and has made a plea for the publication of more measured coefficients, with precise descriptions of the conditions under which they were measured. However, it is likely that wear prediction in lubricated sliding will rely on realistic laboratory tests and tests under field conditions for the foreseeable future.

One particularly damaging type of wear under lubricated conditions is *scuffing*, a form of wear in which localised solid-phase welding occurs between sliding surfaces, resulting in damage and roughening of the surfaces. The original area of scuffing is almost invariably roughened to such an extent that it is likely to penetrate the oil film on subsequent contacts. Some scuffed areas are self-healing – i.e. they undergo a period of mild wear which results in smoothing of the roughened surface. However, in other cases the problem becomes rather worse on each contact, with both the severity and the extent of roughening increasing. In this latter case the damage can lead very rapidly to gross damage and fracture of the sliding components. Thus, it is characteristic of scuffing failure that it occurs very rapidly: mechanisms which have been running very smoothly can suddenly become noisy, and failure often ensues within minutes.

Scuffing failure is often observed on components which are expected to be separated by hydrodynamic or elastohydrodynamic films, particularly important examples being piston rings and gear teeth. The oils used in such situations always include boundary lubricants, which are expected to provide protection on the occasions when the full-film protection breaks down. It is, therefore, clear that the onset of scuffing is associated with the breakdown of these boundary lubricants, but the factors which cause this breakdown are by no means clear.

The subject has been admirably reviewed by Dyson (1975), who states that the onset of scuffing is probably associated with the attainment of a critical temperature at the contacts. However, Dyson states that the questions which remain to be answered include: how scuffing is initiated; how initiation proceeds to propagation; whether lubricants control scuffing; which chemical reaction products are important; and what the functionally important characteristics of the reaction products are. As Dyson concludes, these questions give a good summary of the present state of ignorance of the subject.

## 3.10 PROBLEMS

1 In a wear test, a bronze annulus having an outside diameter of 25 mm and an inside diameter of 15 mm is placed with its flat face resting on a flat carbon steel plate under a normal load of 100 N and rotated about its axis at 5 Hz for 20 h. At the end of the test, the specimens are separated and weighed and it is found that the mass losses of the bronze and steel are 250 mg and 8 mg, respectively. Using the materials data given below, calculate wear coefficients for the bronze and the steel.
(Hardnesses (MPa): steel, 2400; bronze, 800.
Densities ($Mg/m^3$): steel, 7.8; bronze, 8.4.)
$[K_{bronze} = 7.4 \times 10^{-4}; K_{steel} = 6.57 \times 10^{-5}]$

# Chapter 4
# Tribological Properties of Solid Materials

## 4.1 METALS

Much of the material in the previous chapters has been concerned with the tribological properties of metals, and it is not necessary to reiterate it here. In particular, the effect of hardness on resistance to wear was mentioned frequently in Chapter 3 and the importance of oxide films was stressed in Section 3.5. However, in covering the basic mechanisms of wear some other properties which can have a marked effect on wear behaviour were mentioned only in passing, or not at all. The effects of these properties are described in the following sections.

## 4.1.1 Work Hardening and Ductility

One effect of work-hardening characteristics was mentioned in Section 3.4.1, where it was stated that the resistance to milder forms of abrasive wear can increase with the work-hardening capacity of a metal. However, work hardening combined with high ductility can lead to very poor tribological properties. This is because high ductility can allow plastically deforming asperities to undergo large amounts of plastic deformation before fracture. If the plastically deforming asperities are work hardened to a hardness level considerably above that of the opposing surface or, in the case of transferred material, above that of the parent metal, they can then cause severe abrasive wear. Furthermore, high levels of plastic deformation almost invariably cause disruption of protective oxide films, thus leading to strong interfacial adhesion and exacerbation of the situation described above. Therefore, materials which combine high ductility with the ability to work harden often exhibit both high friction and severe adhesion-enhanced wear, where material loss may be slight but damage to the sliding surfaces is very severe.

Important examples of such materials are austenitic stainless steels and alloys of titanium. These materials have poor tribological properties in many situations, particularly where the loads are very high. For this reason, the materials are notoriously difficult to machine, owing to pick-up (adhesive transfer of the material to the cutting tool) and consequent blunting of the tool and damage to the workpiece. However, the austenitic steels may have good tribological properties under lighter loads, particularly at high temperatures in oxidising environments. This is because such steels naturally form tenacious, protective films of chromium oxide and, if the loads do not cause disruption of the films, can thus provide an excellent sliding surface.

## 4.1.2   Effect of Crystal Structure

In Chapter 3 we implicitly treated metals as isotropic materials which would yield and plastically deform at a certain stress level. We took no account of anisotropy (i.e. the variation of mechanical properties with crystallographic direction) or of the direction of the stress axis relative to the crystal axes.

These simplifications do not introduce any significant error in describing the tribological behaviour of many common metals, particularly those with face-centred cubic (FCC) and body-centred cubic (BCC) crystal structures. These materials are almost isotropic and have a multiplicity of slip systems, which allows them to yield at similar stress levels, whatever the crystallographic orientation of the stress axis. (Readers who are unfamiliar with the terminology used here should consult one of the standard texts on the plastic deformation of metals, such as Honeycombe, 1968.)

However, there is one important class of metals for which these assumptions are far from justified, and this leads to important, and usually beneficial, consequences for their tribological properties. These metals are those with close-packed hexagonal (CPH) crystal structures which deform predominantly by basal plane slip, the major examples being cobalt and rhenium.

As these metals have only one slip plane, they cannot deform by slip when the stress axis is either perpendicular or parallel to that plane, and can thus support very high loads elastically. Furthermore, even when the stress axis is oriented so that slip can occur, the anisotropy precludes the isotropic plastic deformation which is necessary for junction growth. Thus, such materials can have acceptably low friction coefficients, even in ultra-high vacuum, when their surfaces are perfectly clean and there is no possibility of the gross seizure which can occur with more isotropic metals in such conditions.

The effect of crystal structure is demonstrated very clearly by the example of cobalt, which undergoes an allotropic change from CPH to FCC at 417°C. As shown in Figure 4.1, at the point where this transformation occurs both the friction coefficient and the wear rate increase dramatically. (Note that

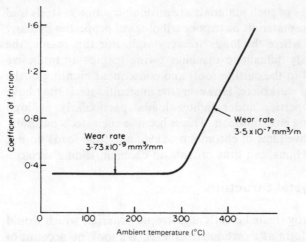

**Figure 4.1** Friction coefficient and wear rate for cobalt on cobalt in vacuum at various temperatures: pressure, $10^{-7}$ N/m$^2$; sliding speed, 2 m/s; load, 9.81 N

in Figure 4.1 the ambient temperature at the transition is less than the quoted temperature of 417 °C; this is because frictional heating causes the actual surface temperature to be higher than the ambient temperature.) Fortunately, this allotropic transformation temperature can be increased by additions of suitable alloying elements. For this reason many wear-resistant materials are based on cobalt–30% chromium, which retains the CPH structure up to a temperature of $> 800$ °C.

### 4.1.3  Effects of Microstructure

In Chapter 3, where we described the basic mechanisms of wear, we largely confined our attention to homogeneous, single-phase materials. We did this because, with such materials, each asperity contact will be between the same two materials and the mechanical properties of all the asperities can be assumed to be similar to those of the bulk parent materials. This allows us to develop clear descriptions of the effects of properties such as material hardness. However, by adding other elements to a material – i.e. by alloying – it is possible to change material properties in various ways, often with beneficial effects on wear behaviour. It would be beyond the scope of this book to detail the effects of alloying, but for an excellent introduction to the subject the reader is referred to Ashby and Jones (1986). However, for our purposes, we can divide the mechanisms into two groups: those resulting in single-phase alloys, and those resulting in multiphase alloys.

There are two principal effects which can cause a single-phase alloy to be more wear-resistant than the parent metal: solution hardening and transformation hardening. These effects have recently been reviewed by

Glaeser (1980) and (particularly for the technologically very important ferrous metals) by Borik (1980). The more important of the two effects is transformation hardening, particularly in the heat treatment of steels to cause a martensitic transformation. This transformation can lead to increase in hardness of almost an order of magnitude and a correspondingly great reduction in ductility, both of which will cause improvements in wear behaviour. However, the brittleness of martensitic materials means that they are prone to tensile failure. Therefore, they must either be used in compression or tempered to improve their toughness, at the expense of some reduction in hardness and increase in ductility.

Multiphase alloys for wear-resistant applications usually consist of a relatively soft matrix containing a fine dispersion of relatively hard particles. This type of structure can confer two principal beneficial effects on tribological properties. First, the fine dispersion can limit the area over which junction growth can occur and can thus reduce both friction and adhesion-enhanced wear; for this reason it is a rule of thumb that the improvement in tribological properties increases with the fineness and concentration of the dispersed second phase. Second, the hard particles can confer excellent abrasion resistance when they are harder than any potentially abrasive particles in the environment. However, they can significantly worsen the situation if they are softer than the external abrasive, as they can be loosened from the parent matrix and cause additional abrasive wear.

The cobalt-based superalloys combine the crystal structure effect described in sub-section 4.2.2 with precipitation effects due to a fine dispersion of hard carbides; these materials have excellent tribological properties in a wide range of applications.

### 4.1.4 Mutual Solubility of Rubbing Pairs

Rabinowicz (1970) has suggested that pairs of materials with a high degree of mutual solubility will have poor tribological properties, as a consequence of high interfacial adhesion resulting from their mutual interdiffusion and bonding. He attempted to correlate mutual solubility, as indicated by binary alloy equilibrium diagrams, with adhesion, friction and wear. The analysis showed zero correlation with adhesion; some positive correlation with friction; and a greater correlation with wear. This suggests that pairs of metals for boundary lubricated or unlubricated sliding should be chosen to have low mutual solubilities. This is a generally accepted rule of thumb but, as with most tribological rules, there are some notable exceptions. For example, the rule suggests that the same material should not be specified for each component of a rubbing pair, but some materials, particularly the cobalt-based superalloys mentioned above, have excellent tribological properties against themselves. Recently Tabor (1985), Buckley (1985) and Ferrante et al. (1985) have shown that the situation is much more complex than is suggested by this simple rule, and that further understanding of interfacial adhesion will need to be based on a better understanding of such factors as the electronic structure of the interface and the degree of misfit between the crystal lattices of the two surfaces. It is worth noting that two lattices with a high degree of misfit are likely to have low mutual solubilities, so it may be difficult to separate these effects.

### 4.1.5  Effects of Temperature on the Above Properties

It is clear that a change of temperature could have a profound effect on most of the properties described above. As the temperature is raised, any single-phase materials, including those with martensitic structures, will tend to become softer and more ductile; rates of oxidation will increase and the type of oxide formed may change; hard precipitates may be dissolved, or grow and become less effective; crystal structures may change; and mutual solubility may increase. Therefore, it is obvious that, when choosing rubbing pairs, it is necessary to choose them on the basis of the material properties obtaining at the predicted operating temperature. As an example, for operation at elevated temperatures it is usually necessary to specify materials with high hot hardness, such as tool steels or superalloys.

## 4.2  BEARING ALLOYS

We use the term 'bearing alloys' to describe alloys developed for use in well-lubricated journal bearings. As will be shown in Chapter 6, under normal running the bearing and its counterface – often a crankshaft – are completely separated by a thin oil film which is generated by hydrodynamic lift. Under such conditions the tribological properties of the solid material are, of course, irrelevant. However, the oil film is not present for a short time after each start of the engine, or when the engine is running slowly at high loads. Under these circumstances the surfaces are protected only by boundary lubrication, as described in Chapter 2, and as boundary lubrication involves some solid–solid contact and high asperity stresses, the tribological properties of the materials are then very important.

### 4.2.1  Required Properties of Bearing Alloys

When we use bearing alloys, we are not usually in the fortunate position of being simply concerned with two chosen materials rubbing together in the presence of a film of clean oil containing boundary lubricants. For example, the inside of an internal combustion engine is a very hostile environment, containing corrosive combustion products, abrasive wear debris and usually – particularly in new engines – cast-iron dust left behind after machining the engine block, and wear debris from the piston, cylinder and piston rings formed during running in. The rubbing materials must, therefore, be able to withstand most of the wear processes described in Chapter 3. Furthermore, the bearing must sustain very high fluctuating dynamic loads, which are inherent in the operation of an engine, without suffering fatigue failure. These various, and often conflicting, requirements have led to the development of ranges of bearings which often contain several components. The compositions, properties and methods of manufacture of the various types of alloy cannot be described in detail here and the reader requiring more detail should see, for example, the excellent review by Morris (1967) or *The Tribology Handbook* (Neale, 1973). We shall simply describe the general requirements of such bearings and briefly mention some examples. The required characteristics are described in turn below.

## Embeddability

As mentioned above, engines often contain abrasive particles, and, if these are larger than the oil film thickness at its thinnest, they cannot pass through the bearing. Thus, unless both the bearing and the crankshaft are considerably harder than the abrasive particles, there is the potential for abrasive wear to occur. This problem is solved by making the surface of the crankshaft hard enough to resist abrasion (using one of the processes described below in Section 4.7) and making the bearing alloy so soft that the abrasive particles can be pushed into the bearing surface, where they are permanently trapped and rendered harmless. This ability of the alloy to trap abrasives is known as embeddability.

The reader may wonder why the abrasion problem is not solved by making *both* rubbing surfaces harder than the abrasive. This is explained by the need for other characteristics, as described below.

## Conformability

It would be extremely difficult and expensive to manufacture a crankshaft and a set of bearings with no significant misalignments between the bearings. This problem is also solved by making the bearing alloy so soft that it can deform plastically on assembly and early running, and thus accommodate any initial misalignments. This attribute of the bearing is known as conformability.

## Protection of the Crankshaft

A crankshaft is a relatively expensive component to manufacture and install, compared with a bearing shell, and the crankshaft–bearing system should, therefore, be designed so that wear is largely confined to the bearing shell, with the crankshaft remaining relatively unscathed. It should now be clear from reading Chapter 3 that this is best accomplished by making the rubbing surface of the bearing shell much softer than that of the crankshaft. Furthermore, the soft metals chosen for the bearing surface will also have low melting points, so in the extreme case of complete oil starvation the surface of the bearing can melt and lubricate the rubbing surfaces, thus preventing catastrophic engine seizure.

### Load-carrying Capacity and Fatigue Strength

So far, all our requirements have led to the choice of a soft material for the bearing shell and, at first sight, this conflicts with the need for the bearing to carry very high fluctuating loads without gross plastic deformation or fatigue failure. However, we can have all the properties listed above *and* good mechanical strength by making the soft metal very thin. The reason for this is that when a very thin film of soft metal is supported on a hard base, it can carry very high normal pressures without yielding. For example, El-Shafei *et al.* (1983) have shown that thin lead films on hard steel substrates can sustain the same pressure as the uncoated steel before yielding. This is because the film is trapped between two extensive hard surfaces and it can deform plastically only by flowing outwards so that some material leaves the contact zone. As the film becomes thinner, the pressure required to cause this outward flow increases very rapidly, and tends to infinity at zero film thickness. This effect of plastic constraint is used in bearing design by supporting a thin and soft bearing material on a relatively strong and hard steel backing strip.

### 4.2.2  Compositions and Structures of Bearing Alloys

The two metals which combine best the required properties of softness, low friction against steel and low melting point are tin and lead. However, in order for these materials in the unalloyed form to have the required load-carrying capacity they would need to be no more than a few microns thick, and this would both limit the lifetime and restrict the embeddability and conformability of the bearing. These metals are therefore universally used in two types of alloy: first, as lead-based or tin-based alloys, in which properties such as hardness and strength are enhanced considerably by small alloying additions; second, as dispersions in a matrix of a stronger material. The two most common materials of this second type are copper–lead and aluminium–tin alloys; in each of these the zones of the softer material interconnect, so that any lead or tin lost from the rubbing surface can be continuously replenished from the underlying material. Typical microstructures of alloys of such materials are shown in Figure 4.2.

The copper–lead and aluminium–tin alloys are stronger and harder than those based principally on lead and tin. However, the lead component

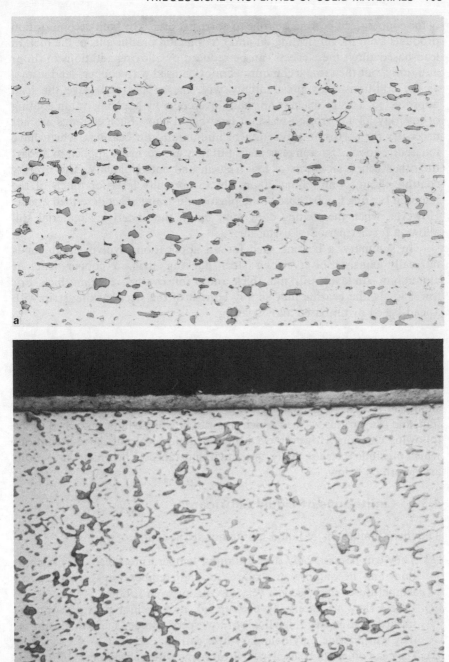

**Figure 4.2** (a) Microstructure of an aluminium–tin-based bearing alloy (courtesy, Glacier Metal Co. Ltd, Kilmarnock); (b) microstructure of a cast copper–lead bearing with an electroplated lead–indium overlay (courtesy, Vandervell Ltd, Maidenhead)

of the copper–lead bearing is prone to corrosive attack by both decomposition products from the lubricating oil and combustion products from the fuel. In lead-based alloys these effects can be reduced by alloying additions of tin or antimony, but the same techniques cannot be used with copper–lead alloys, as the additions alloy preferentially with the copper and leave the lead unprotected. This problem is normally overcome by applying a thin coating or 'overlay' of a corrosion-resistant lead-based alloy. Such coatings are not necessary for corrosion protection of aluminium–tin bearings, which have inherently good corrosion resistance, but they are normally applied to impart the other beneficial effects of low friction and embeddability. The overlay coating, though only about 30 $\mu$m thick, normally lasts for much of the required bearing life, because of the plastic constraint effect described above. However, when the coating does wear away, the underlying material still allows a further period of useful life.

In summary, a bearing shell for an application involving high loads normally consists of a strong steel backing strip, a copper–lead or aluminium–tin layer $\approx 0.4$ mm thick and a soft overlay $\approx 30$ $\mu$m thick. For lower loads, the bearing may simply consist of the steel backing strip with a lead-based or tin-based alloy coating.

## 4.3  SOLID LUBRICANTS

There is no doubt that the ideal way to reduce both friction and wear of sliding surfaces is to arrange for them to be completely separated by a fluid film, and the various ways in which this can be done will be described in the following chapters. However, even the best fluid lubricants have limitations which preclude their use under certain circumstances. Their main limitations are described below.

### 4.3.1  Limitations of Liquid Lubricants

The main limitations of liquid lubricants are as follows.

(1) They can be used only over a limited temperature range; they become thin, and decompose or oxidise, at high temperatures, while at low temperatures they become very viscous or even solidify.
(2) Unless they are externally pressurised, they cannot provide complete surface separation at high loads or low speeds; under such circumstances they can provide only boundary lubrication.
(3) They can often not be used in environments where cleanliness is essential, such as in food processing equipment.
(4) They are prone to attack in hostile environments such as acids and solvents.

(5) They cannot be used to provide permanent lubrication for mechanisms which are inaccessible after assembly.

(6) They are degraded by radioactive environments.

These limitations can often be overcome by using solid lubricants. Furthermore, when solid lubricants can be used, they remove the need for expensive lubricating systems and they can greatly reduce the need for periodic maintenance.

### 4.3.2 Requirements of Solid Lubricants

The basic requirement of any solid lubricant is exactly the same as that for a liquid lubricant. It should completely separate two solid surfaces, even under a very high normal load, while allowing the surfaces to slide over each other under a relatively low shearing force: in other words, a solid lubricant must meet the twin requirements of having a high normal load capacity and a low shear strength.

We have already described one type of solid lubricant in sub-section 4.2.2, where we showed that these twin requirements can be met by a thin film of a soft metal supported on a relatively hard substrate. However, this type of solid lubrication is a property of the multicomponent *system*, rather than an inherent property of a single material, and in this section we intend to describe single materials which can act as solid lubricants.

It is clear that the twin requirements described above are unlikely to be met by any isotropic material, in which the shear strength will be about 16% of the yield pressure, and it has been found that the best solid lubricants are very anisotropic materials known as lamellar solids.

### 4.3.3 Lamellar Solids

A lamellar solid is one in which the atoms are bonded together in parallel and comparatively widely spaced sheets, the two best-known being graphite and molybdenum disulphide, which have the crystal structures shown in Figures 4.3 and 4.4, respectively. Under many circumstances these are excellent lubricants, as are other less well-known lamellar solids such as tungsten diselenide and cadmium iodide. However, not all lamellar solids are good solid lubricants, and there is as yet no theory which will tell us whether or not any particular lamellar solid will be an effective lubricant.

The lamellar solids which can lubricate effectively do have certain characteristics in common. First, they form a strongly adherent transfer film on the surface being lubricated, so that, after the short running-in period during which this film is formed, the actual interface is between lubricant and lubricant. Second, the lubricant on both surfaces develops the preferred orientation shown schematically in Figure 4.5, and this reduces the mechanical

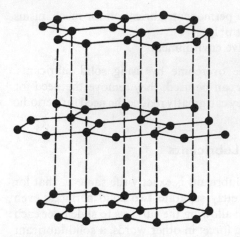

**Figure 4.3** Crystal structure of graphite

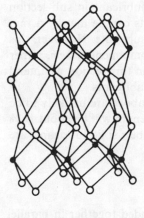

● Molybdenum
○ Sulphur

**Figure 4.4** Crystal structure of molybdenum disulphide

interaction between the surfaces, as can be demonstrated by reversing the direction of motion, when the friction increases significantly.

Although there have been many experimental investigations of the lubricating properties of graphite and molybdenum disulphide, there is no universally accepted explanation of their behaviour.

The earliest explanation of the properties of graphite (Bragg, 1924) was that its shear strength parallel to the atomic sheets was inherently very low. This explanation was accepted until World War II, when it was found that graphite brushes in the electrical generators of high-flying aircraft wore out very rapidly. This effect was investigated by Savage (1948), who showed that

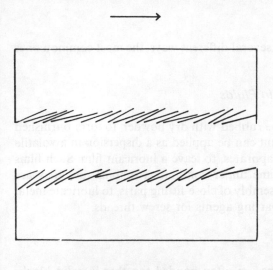

**Figure 4.5** Schematic illustration of the surface orientation developed during the rubbing of lamellar solid lubricants

graphite has very poor tribological properties in the absence of condensable vapours. This effect is very marked and the amount of vapour required to give effective lubrication can be very low. The effectiveness of any vapour seems to depend on the size of its molecules. Nitrogen is ineffective, even at atmospheric pressure; water provides very effective lubrication at a pressure of 400 Pa; and many long-chain hydrocarbons are effective at pressures as low as $\simeq 1$ Pa. The transition in behaviour is startling: at the critical pressure the wear rate can change by about three orders of magnitude and the friction coefficient by a factor of 5. It is now thought that the condensable vapours saturate any available valence bonds at the surfaces and edges of the individual crystallites, so that the adhesion between neighbouring crystallites is very low and they can easily slide over each other. This explanation has been supported by the work of Arnell *et al.* (1968), who have shown that even strong and brittle mechanical carbons and pyrolitic carbon exhibit the same low-friction behaviour as graphite.

Unlike graphite, molybdenum disulphide appears to be an intrinsically good lubricant, and its friction coefficient under high-vacuum conditions is lower than that in air. There are alternative suggested explanations of this inherent low friction. The first is that it is a true low-shear-strength material, with shear taking place between the neighbouring layers of sulphur atoms shown in Figure 4.4. The second, which is similar to that now accepted for graphite, is that any available surface and edge valences are saturated by reaction with oxygen atoms but, unlike the vapours on graphite, the oxygen is not volatile and persists up to the decomposition temperature of the solid.

### 4.3.4 Methods of Use

Solid lubricants can be used in several different ways, the most common ones being described below.

#### Dry Powders or Dispersions in Fluids

Surfaces to be lubricated can be rubbed with dry powder, to form burnished films. Alternatively, the lubricant can be applied as a dispersion in a volatile solvent which subsequently evaporates, to leave a lubricant film. Such films would have only a short lifetime under continuous rubbing, but they are used extensively to facilitate assembly of close-fitting parts, to lubricate metal working components and as parting agents for screw threads.

#### Solid Blocks

Graphite and graphitic carbons are often bonded together in solid blocks which can, for example, be made into solid thrust bearings and journal bearings. The blocks are usually made from a mixture of finely divided coke and a carbonaceous binder such as pitch. This mixture is fired to a very high temperature, to cause it to graphitise, and by varying the heat treatment a wide range of materials can be made. These materials range from highly crystalline electrographite, which is generally used for low-load applications such as electrical generator brushes, to almost amorphous mechanical carbons, which are used for bearings.

#### Additives to Oils and Greases

The major use of molybdenum disulphide is probably as an additive to oils and greases. In this form it is very beneficial during running-in operations and it increases the effectiveness of lubrication under heavy loads. When used in engine oil, it is deposited as a thin solid film at the piston–cylinder interface, and this can offer some short-term protection in the event of complete oil loss. If such additives are used in lubricating systems containing a filter, it is essential that the particle size be sufficiently small to allow the particles to pass through the filter, which would otherwise quickly become clogged.

#### Resin-bonded Films

Solid lubricants are frequently bonded to metal surfaces by use of resin binders, and such films can have wear lives some orders of magnitude higher than those of dry powder films. The only drawback of these films, compared with dry powder films, is that thermal decomposition of the resin imposes

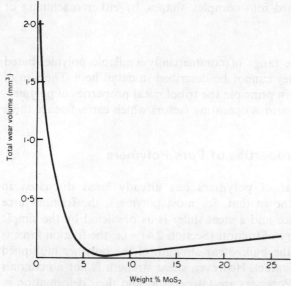

**Figure 4.6** Wear of powder metallurgy specimen containing molybdenum disulphide

a limit on the maximum rubbing temperature, and therefore on the load and sliding speed.

### Metal-based Composites

The thermal limit mentioned in the previous paragraph can be overcome by incorporating the lubricant in a metal matrix. The relative proportions of matrix and lubricant are quite critical, as can be seen from Figure 4.6. This figure illustrates the necessity of careful control of materials content if the twin requirements of high load-carrying capacity and low shear strength are to be met. If the composite carries too little lubricant, it has a very high load-carrying capacity but also a high shear strength, whereas, if it carries too much lubricant, it has a low shear strength but also a reduced load-carrying capacity.

## 4.4 POLYMERS AND COMPOSITES

Polymer-based bearings can have all the advantages of solid lubricants listed in the introduction to Section 4.3, plus the following additional advantages.

(1) They absorb vibrations well and are quiet in operation.
(2) They readily deform to conform to mating parts, so that machining tolerances and accuracy of alignment are usually less critical than for metal parts.

(3) They are easily formed into complex shapes, by either machining or moulding.

(4) They are cheap.

There is a very wide range of commercially available polymer-based bearing materials, and they cannot be described in detail here. The aim of this section is to describe in principle the tribological properties of polymer-based materials and the various operating factors which can influence these properties.

### 4.4.1  Tribological Properties of Pure Polymers

The frictional behaviour of polymers has already been discussed in sub-section 2.6.2. It is known that, for most polymers, the friction force between a polymer surface and a steel slider is as predicted by the simple Bowden and Tabor theory of friction (Section 2.4) – i.e. the friction force is approximately equal to the bulk shear strength of the polymer multiplied by the effective area of contact. However, shear strength is not a materials constant for a polymer. Polymers are viscoelastic, and their deformation is, therefore, dependent on strain rate; consequently, friction forces can vary with such parameters as sliding speed and surface roughness.

However, the friction coefficients of most polymers, against metals and against themselves, are usually in the range 0.2–0.4. Polytetrafluoroethylene (PTFE) is an outstanding exception: the friction coefficient for PTFE sliding on itself can be as low as 0.04, which is the lowest known value for any solid. There is no firm explanation for this phenomenon, although it is usually ascribed to the intrinsically low chemical bonding forces between PTFE molecules, which consist of relatively stiff chains of carbon atoms surrounded by larger fluorine atoms. Therefore, molecules at the surfaces can slide over each other under very low shear forces, while mechanical interlocking of the molecules in the bulk material gives it a relatively large load-carrying capacity and bulk shear strength. As already described in sub-section 2.6.2, when PTFE slides against a counterface of normal roughness, a thin film of the polymer is transferred to the counterface, so that the observed friction coefficient is due to the polymer sliding on itself.

We can see that, with the exception of PTFE, polymers do not have particularly low coefficients of friction. The tribological property of polymers which causes them to be so widely used is that they wear at low and reasonably predictable rates, thus allowing a designer to select with some confidence a bearing which will give the required operating life under specified conditions of load and speed. We shall return to this subject in sub-section 4.4.3.

### 4.4.2  Effects of Reinforcements and Fillers

There are only five basic polymers in common use in tribology: PTFE, polyacetal, Nylon, polyethylene and polyimide, although other materials are now coming into use. However, many reinforcements and fillers have been incorporated into these basic materials, so that a wide range of bearings is commercially available.

As their name implies, the purpose of reinforcements is to increase the mechanical strength of the bearing. Reinforcement is carried out in one of three basic ways. First, a mechanically strong, normally fibrous material is mixed with the polymer before moulding, the most common example being chopped glass fibre, which increases strength, stiffness and creep resistance. Second, the polymer is incorporated into a woven matrix; with thermosetting resins the woven matrix is soaked in the resin, which is then allowed to harden, whereas with thermoplastic polymers, such as PTFE, the polymer fibre and reinforcing fibre are woven together in such a way that the actual bearing surface is predominantly the lubricating polymer. Finally, the polymer can be supported as a relatively thin layer on a metal backing plate, to give the plastic constraint and increased load-bearing capacity already described for bearing alloys in sub-section 4.2.2. If the polymer is also supported around its edges, so that it is effectively held in a metal container, its load-carrying capacity can be increased enormously. Pure PTFE has very poor creep resistance and, if unconstrained, will creep under very low loads, but if fully constrained, as described above, it can be used to carry very high loads – for example, in civil engineering applications such as bridge bearings.

A metal liner on a thin polymer bearing has the additional advantage of greatly assisting heat dissipation from the sliding interface. In general, polymers have low thermal conductivities and high thermal expansion coefficients when compared with metals, and, therefore, if solid bushes of pure polymer are used against steel journals, frictional heating can cause bearing seizure. To avoid this, it would be necessary to make the polymer bearing to a much looser fit than the designer would otherwise wish. However, the problems of low conductivity and high expansion coefficient can both be avoided by using the polymer as a thin film on a metal liner.

Fillers, which are particulate materials, are used to improve strength, thermal properties and frictional behaviour. Metal powders of high conductivity and relatively high strength are used to improve thermal and mechanical properties, while graphite, molybdenum disulphide and soft metals such as lead are used to improve frictional behaviour.

### 4.4.3 The PV Factor

The rate of wear of a polymer will obviously be some function of the load and sliding speed, and it is found that we can define for any polymer bearing a design criterion known as the $PV$ factor.

The theoretical derivation of the relationship between the wear rate and the $PV$ factor starts from the reasonable assumption that the rate of wear will be proportional to the rate of energy dissipation at the sliding interface. On this basis, we can then derive a relationship between the rate of wear

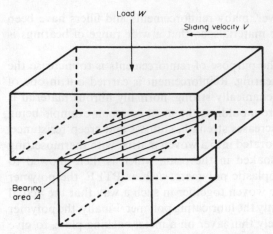

**Figure 4.7** Schematic diagram of flat bearing for derivation of $PV$ factor

and the $PV$ factor for the two basic bearing configurations.

(1) *Flat bearings* If we have a flat bearing surface of area $A$, as shown in Figure 4.7, subjected to a normal load $W$ and sliding against its counterface under a coefficient of friction $\mu$, then the energy dissipated in sliding through a distance $dx$ is given by $\mu W \cdot dx$, and the rate of energy dissipation is given by $\mu W(dx/dt) = \mu W V$, where $V$ is the sliding speed at the interface.

We can then state our assumption, that the volume wear rate, $Q$, is proportional to the rate of energy dissipation, as

$$Q = kWV \qquad (4.1)$$

where the coefficient of friction, $\mu$, which, for the purpose of this derivation, we have assumed to be constant, has been subsumed into the constant of proportionality, $k$.

By rearranging Equation (4.1), we can see that the constant of proportionality, $k$, is the volume wear per unit load per unit slid distance; therefore, $k$ is known as the specific wear rate of the material.

We are normally interested in the *depth* of wear, or the rate of linear wear normal to the sliding surface, and we can see that the depth of wear at any time is simply the volume of wear divided by the bearing area. Therefore, the rate of linear wear, $Q_d$, is given by $Q/A$, so that

$$Q_d = kWV/A = kPV \qquad (4.2)$$

where $P$ is the apparent pressure on the bearing.

(2) *Journal bearings* Figure 4.8(a) shows a journal bearing supporting a shaft, under a normal load $W$, on the shaded bottom half of the bearing. We assume that we have a normal reaction, $R$, per unit area of the load-

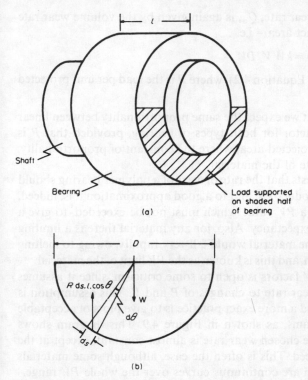

**Figure 4.8**   Schematic diagram of solid lubricated journal bearing and forces acting

bearing surface. By considering the equilibrium of vertical forces, we see that the normal load $W$ must be kept in equilibrium by the vertical component of this reaction integrated over the bearing area. From Figure 4.8(b) we can see that the reactive force over an element of surface of length d$s$ will be given by $Rl \cdot \mathrm{d}s$, where $l$ is the axial length of the bearing, and the vertical component of this reaction will be $Rl \cos \theta \cdot \mathrm{d}s$. But $\mathrm{d}s = D \cdot \mathrm{d}\theta/2$, so that the total vertical reaction is given by

$$(Rl D/2) \int_{-\pi/2}^{\pi/2} \cos \theta \, \mathrm{d}\theta = Rl D$$

Therefore $R = w/Dl$. But the rate of energy dissipation per unit area of interface is $\mu PV$, so that the total rate of energy dissipation is given by

$$\mu RV \cdot \pi D l/2$$

$$= (\mu W/Dl) \cdot V \cdot (\pi Dl/2)$$

$$= \mu W V \pi/2$$

Therefore, the volume wear rate, $Q$, is given by $Q = kWV\pi/2$, where, as before, $\mu$ has again been subsumed into the constant.

But the radial wear rate, $Q_d$, is again given by the volume wear rate divided by the contact area – i.e.

$$Q_d = Q/(\pi Dl/2) = kWV/Dl$$

so that $Q_d = kPV$ (cf. Equation 4.2), where $P$ is the load per unit projected area.

Thus, we can see that we expect the same proportionality between linear wear rate and the $PV$ factor for both types of bearing, provided that $P$ is taken as load per unit projected area, where the constant of proportionality, $k$, is the specific wear rate of the material.

Equation (4.2) suggests that the rate of wear of a polymer bearing should be proportional to the product $PV$, and to a good approximation it is, indeed, often possible to specify a $PV$ value which must not be exceeded, to give a certain wear rate or life expectancy. Also, for any material there is a limiting $PV$ factor above which the material would fall very rapidly, owing to melting or thermal decomposition, and this is known as the $PV$ limit of the material.

The use of single $PV$ factors is open to some criticism, since it assumes the same sensitivity of wear rate to changes of $P$ and $V$. This assumption is frequently unfounded, and a more exact practice is to give plots of acceptable wear rates on $PV$ diagrams, as shown in Figure 4.9. This diagram shows that the $PV$ factor for the chosen wear rate is almost constant, except at the extremes of load and speed. This is often the case, although some materials have $PV$ diagrams which are continuous curves over the whole $PV$ range.

It is also found that there is a limited range of low pressures and temperatures over which $k$ can be taken to be constant, and within this region $k$ is given the symbol $k_0$. Thus, within its range of applicability, Equation (4.2) can be used to predict wear rates; it is only necessary to measure the wear rate at one value of $PV$ to determine the value of $k_0$, and this value can then be used to predict wear under other $PV$ conditions.

At higher temperatures or pressures the value of $k$ increases, and for accurate data recourse must be had to detailed design manuals. In particular, the ESDU *Guide to the Design and Material Selection for Dry Rubbing Bearings* (1987) should be regarded as an indispensable guide to any designer involved in the specification of such components. This guide gives not only $PV$ data of the type described above, but also detailed information on the effects of other variables such as the friction coefficient, which we initially assumed to be constant, the counterface composition and roughness, the presence of lubricants, and the thermal properties of the bearing assembly.

The $PV$ factors or $PV$ curves which are published or supplied by bearing manufacturers normally refer to unlubricated sliding against a counterface of a specified material having a specified surface finish. When the sliding conditions depart from these standard conditions, the designer must take into account such changes when specifying a bearing, and to do this it is essential that a more detailed guide, such as the ESDU guide, be consulted.

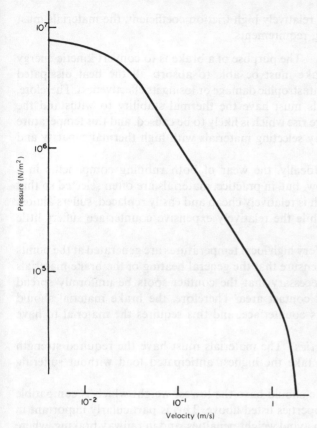

**Figure 4.9** Typical limiting $PV$ curve of PTFE-based material for wear rates of 25 $\mu$m in 100 h

## 4.5 Materials for Brakes and Clutches

The materials used in the manufacture of brakes and clutches are known as friction materials. Braking is usually by far the more demanding of these applications, and in what follows we shall concentrate on brakes, although the same principles will also apply to clutches.

In contrast to all other tribological applications, high friction coefficients are desirable when using friction materials, and the materials are normally developed to give friction coefficients, against the chosen counterface, of the order of 0.35–0.55. The principal frictional requirement is that the friction coefficient should be stable with respect to sliding speed, time and temperature, to avoid the brakes fading during a prolonged application, but a high friction coefficient ensures that the required braking force can be obtained under a relatively low normal load.

In addition to the relatively high friction coefficient, the materials must also meet the following requirements.

(1) *Thermal properties*  The purpose of a brake is to convert kinetic energy into heat. The brake must be able to absorb all the heat dissipated without suffering catastrophic damage or losing its effectiveness. Therefore, the brake materials must have the thermal stability to withstand the highest temperature rise which is likely to be caused, and this temperature rise is minimised by selecting materials with high thermal capacity and conductivity.

(2) *Wear resistance*  Ideally, the wear of both rubbing components in a brake should be low, but, in practice, materials are often selected so that one material, which is relatively cheap and easily replaced, suffers almost all of the wear, while the relatively expensive counterface suffers little damage.

(3) *Conformability*  Very high local temperatures are generated at the points of contact, and to ensure that the general heating of the brake materials is uniform, it is necessary that the contact spots be uniformly spread over the apparent contact area. Therefore, the brake material should conform well to its counterface, and this requires the material to have a low modulus.

(4) *Mechanical properties*  The materials must have the required strength and toughness to take the highest anticipated load without suffering mechanical failure.

(5) *Weight*  The brake should have the lowest weight which is compatible with the other properties listed above. This is particularly important in aircraft braking, to avoid weight penalties, and in railway braking, where the brakes are usually part of the unsprung mass, to ensure a smooth ride.

By far the most common use of friction materials is in the brakes of road vehicles. These comprise a polymer-based composite friction material rubbing against a cast-iron counterface. The friction material component, the brake shoe or brake pad, is cheap and easily replaceable and is normally replaced several times during the life of the vehicle, while the counterface, the drum or disc, is normally not replaced.

The friction material is a multicomponent composite with a formulation which is developed empirically. The material will normally comprise:

A matrix, of a thermosetting polymer, which confers conformability.

A fibrous reinforcement, to give the required strength and thermal stability. This has traditionally been asbestos but, for health and environmental reasons, this is now being superseded by alternative materials such as chopped steel wool.

A filler, to give frictional stability, usually barium sulphate.

High-temperature lubricants, which are designed to melt and lubricate hot spots, examples being lead sulphide, zinc and aluminium.

A scavenger, which keeps the counterface free of transferred material, usually brass chippings.

Such materials are in a constant state of development and it is not uncommon to find friction materials comprising more than ten components.

In railway braking the traditional method has employed a cast-iron shoe rubbing on the steel tyre of the wheel. However, there has been a tendency in recent years to use composite materials, of the type described above, in preference to cast-iron. In either case the demands on the brakes are so severe that it is normal to replace the brake shoes after some hundreds of miles, rather than the tens of thousands of miles which are expected in road vehicles.

The most demanding area of application for friction materials is in aircraft braking. For example, to abort a take-off of a modern airliner, the brakes would be expected to dissipate all the kinetic energy of the aircraft, perhaps 100 MJ, in about 30 s. Because of the short time involved, very little of this heat can be lost from the brake, and the materials must be capable of absorbing it all. Therefore, polymer-based materials are out of the question for this application. Most modern aircraft use multiple brakes consisting of alternate plates of steel and sintered iron. However, there is a large weight penalty involved in the use of such materials, because the large thermal mass required means that the brakes must have a mass of $\simeq 1$ t. To reduce this mass, friction materials are now being manufactured from a carbon–carbon composite. The use of this new material can reduce the weight of the brakes by about 40%, and can increase the replacement intervals. Nevertheless, the materials are still very expensive and they are still used only where weight saving has a very high priority.

## 4.6 CERAMICS AND CERMETS

The present state of knowledge on the friction and adhesion of ceramics has already been discussed in sub-section 2.6.3, where it was made clear that, as yet, little is known of the nature of interatomic bonding across ceramic–metal and ceramic–ceramic interfaces. This is because, particularly with covalently bonded ceramics, it is unclear how many valences are fully saturated and how many are free to form interfacial bonds. However, it has been observed that friction coefficients for ceramic–ceramic sliding are not particularly low, being of the same order as those between metals in normal atmospheres.

On the other hand, it is well known that ceramics can exhibit very low wear rates, compared with metals, and this property is causing ceramics to be used increasingly in tribological situations. Furthermore, ceramics are refractory materials which retain qualitatively similar properties up to high temperatures, and they are also chemically very unreactive, so there is no doubt that they will find increasing use in hostile environments, particularly those involving high temperatures.

We shall discuss specific examples of the uses of ceramics a little later in this section, but before doing that it is useful to discuss the reasons for the good wear behaviour. This is best done by referring again to the plasticity index, $\psi$, where

$$\psi = (E^*/H) \cdot (\sigma/\beta)^{\frac{1}{2}} \tag{1.28}$$

As we stated in sub-section 3.4.2, surface contact between metals will be almost entirely elastic when $\psi < 0.6$ and almost entirely plastic when $\psi > 1.0$. However, while ceramics have similar elastic moduli ($E$) to those of metals, they have very much higher hardness values ($H$). Thus, for similar surface finishes (represented by $\sigma$ and $\beta$) ceramics are much more likely than metals to be subjected to fully elastic surface contact.

To clarify this point, it is useful to consider the ratio $H/E$ from a different standpoint. As $H$ is a simple multiple of the yield stress of a material, this ratio is a measure of the yield *strain* of the material – i.e. the strain which can be accommodated by fully elastic deformation. Thus, materials with a high $H/E$ ratio, such as ceramics, can accommodate much higher deformations elastically than can those with lower $H/E$ ratios, such as metals; this is obviously critically important in determining the mode of deformation as asperities interact.

There is another crucial difference between ceramics and metals: ceramics are brittle, whereas most metals are ductile. Thus, when most metals reach the limit of elastic deformation, they can accommodate additional deformation plastically, whereas ceramics are likely to fracture. It is true that ceramics are capable of undergoing some plastic deformation when there are local hydrostatic stresses, and such conditions often obtain at regions of asperity contact. Nevertheless, if a ceramic asperity is subjected to an increasing load, it is likely to fracture without significant plastic deformation. The higher asperities on a sliding ceramic surface will, of course, be subjected to the highest forces, and if the resulting wear debris can escape from the interface, the surfaces should therefore become progressively smoother, the surface interactions should become more elastic and the wear rate should decrease.

Despite the elastic deformation processes described above, the wear rates of ceramics are never zero and we must ask what causes the residual wear. The first point to note is that we must consider the situation in terms of fracture mechanics. Ceramics are not homogeneous materials and they invariably contain surface cracks and internal flaws. Any flaw above a certain critical size is able to extend if the stress exceeds a critical level, which may be well below the theoretical fracture stress of the material, and thus a short stress cycle may cause a small incremental crack extension. On repeated stress cycling, such incremental extensions can cause the flaw to extend to the point where a loose wear particle is formed. Furthermore, this process is likely to be exacerbated by the presence of a corrosive environment, which causes more interatomic bonds at the tip of the flaw to be broken on each stress cycle.

The propagation of a crack as described above may be responsible not only for the creation of local fractures, but also for large-scale fracture of the bulk material, and it is this which has, until now, restricted the use of bulk ceramics in tribology. However, the situation is now being improved in three ways.

First, ceramics technology is improving rapidly, so that materials which are more homogeneous and have fewer and smaller flaws are now being produced. Such bulk materials will be used increasingly in future and are already being actively investigated for various applications. For example, manufacturers of automobile engines are aiming to design all-silicon nitride engines in which the engine block, pistons and connecting rods will be made from bulk silicon nitride. If a suitable ceramic with the necessary mechanical properties can be developed, the resulting benefits will be: excellent wear resistance at the piston–cylinder interface; the possibility of a higher operating temperature and therefore higher efficiency; and lower inertial forces, owing to the low density of the ceramic.

Second, problems of bulk fracture are being avoided by the use of *cermets* – i.e. materials which consist of mixtures of metal and ceramic. An example of this is metal-bonded tungsten carbide – a mixture which is approximately 90% tungsten carbide with a metal binder. The ceramic grains are coated with the metal, usually cobalt or cobalt–chromium alloy, and are then sintered, to form a dense solid. The metal at the grain boundaries then prevents any crack being propagated across a ceramic–ceramic grain boundary and leads to the composite material being much tougher than a pure ceramic. Such materials are already used extensively as, for example, sealing rings in mechanical seals.

Third, methods have been developed in recent years for depositing ceramic coatings onto metal surfaces; systems of this type will be described in more detail in Section 4.7.

In closing this section, we should note that, even when the stresses are very low and the flaws are unimportant, ceramics, in common with other materials, will still wear at finite rates. It is thought that this residual wear probably takes place on an atom-by-atom basis, as a combination of mechanical, chemical and thermal energies makes it energetically favourable for an atom to leave the surface.

## 4.7   SURFACE TREATMENTS TO REDUCE WEAR

It should now be clear that the tribological behaviour of any component which is not completely separated from its counterface by a fluid film is critically dependent on the properties of the contacting surfaces, and the choice of contacting materials is therefore of paramount importance. However, the material selected for its tribological properties may be very unsuitable for all but the surface region of the component.

We now know that the wear resistance of a surface tends to increase with hardness, and we should therefore choose materials of high hardness, to resist wear. However, high hardness is often associated with brittleness and low impact strength, and these properties may make any hard material which would be chosen for wear resistance unsuitable for the bulk of the component.

Conversely, materials which have low friction coefficients often have low mechanical strength, and a material which would be selected as a self-lubricating solid might be too weak to be used for the bulk of many components.

Finally, materials with good tribological properties are often very expensive and very difficult to fabricate, and the cost of using them to manufacture bulk components could be prohibitive.

These factors lead us to the conclusion that many components to be used in tribological applications should ideally be made from two materials: a bulk material which meets requirements such as strength, ease of manufacture and cost; and a surface material which provides the required tribological behaviour.

We have seen examples of this independent choice of materials earlier in this chapter, in the use of steel backings to support polymer-based bearings, resin-bonded lamellar solid lubricants and soft bearing alloys. In the remainder of this section we shall describe other ways in which these independent choices can be made. We cannot cover the various processes in detail, and readers requiring more information are referred to an excellent review by Wilson (1975) and a more recent paper by Ramalingan (1980). Following the treatment of Wilson, we shall separate the surfacing treatments into two groups: short-term treatments, which are used to assist running in and for short-term processes such as metal forming; and long-term treatments, which are expected to last the full life of the component.

### 4.7.1 Short-term Treatments

Short-term treatments are treatments which interpose a thin, non-welding layer between solid surfaces and thus allow the surfaces to run in without adhesion-enhanced wear or the formation of hot spots which could impair future performance.

Some films which have inherently short lives are self-regenerating. They include: metal oxides; films formed by reaction of the surface with boundary lubricants and EP additives; and graphite and molybdenum disulphide deposited from oils and greases.

Other films are permanently lost after a short time, but during that time they have the required beneficial effect, which may then last the life of the component. The predominant film of this kind is formed by 'phosphating'. Many commercial processes exist for forming layers of metal phosphate crystals on the surfaces of metal, particularly steel and cast-iron, and these films give good corrosion resistance, good short-term wear resistance and good oil retention. The oil retention is particularly good on grey cast-iron surfaces, as the phosphate film forms only on the metal parts of the surface and leaves oil-retaining channels above the graphite flakes.

A second short-term film in very common use is the very thin film of electroplated tin which is used on aluminium pistons to assist running in.

## 4.7.2   Long-term Treatments

Long-term treatments can involve modification of the existing surface, with or without changing its composition, or deposition of a second material onto the existing surface. For clarity, we shall divide the treatments into three groups.

### Surface Treatments without Change of Composition

Steels containing more than $\simeq 0.3\%$ carbon can be hardened by quenching. (Readers requiring more information on the heat treatment of steels are referred for an excellent introduction to Ashby and Jones, 1986, or for a fuller treatment to Honeycombe, 1980.)

For wear resistance combined with toughness and strength, it is usually beneficial to harden only the surface layers of a steel component, so that we have the required combination of a hard, wear-resistant surface on a tough, strong core. This can be accomplished by heating only the surface of the component, either in a flame or by electrical induction, and then quenching. The induction-heating process, in particular, is easily automated and cheap, and is applied to many components such as gears. The process has the additional beneficial effect that the outer hard martensitic layer is left in compression and thus has improved resistance to processes such as fatigue and stress corrosion.

As an alternative to heat treatment, the surfaces of many metals can be hardened mechanically by shot peening – i.e. bombarding with small metal shot at high speed. This is essentially a work-hardening process which also has the advantage of leaving compressive stresses in the surface layers.

### Surface Treatments with Change of Composition

There are two basic ways of changing the composition of the surface layers of a material: by diffusion or implantation.

In diffusion treatments the component is held at high temperature in a specified environment, so that atoms from the environment diffuse into the outer layers of the component. The principal atoms which are used to modify the surfaces of steels are carbon and nitrogen.

Carburisation involves holding low-carbon steel in a carbon-rich environment at high temperature. Carbon from the environment then diffuses into the outer layer and transforms it into high-carbon steel, which can then be heat treated to improve its hardness. Nitriding involves the diffusion of nitrogen into a steel surface to form small crystals of metal nitrides, and nitriding steels usually contain alloying elements which react readily with nitrogen. The nitrogen can be diffused in from gaseous atmospheres, fused salts or argon–nitrogen plasmas (Bell *et al.*, 1983), and nitriding temperatures are usually much lower than those used for carburising. After nitriding, the component often does not require any further heat treatment. Cyaniding and carbonitriding cause both carbon and nitrogen to diffuse into the surface. Cyaniding is a salt-bath process carried out at a nitriding temperature, whereas carbonitriding is a gaseous process carried out at higher temperature.

Many commercial processes exist under particular trade names, for carrying out the processes described above and for diffusing in other elements such as boron, aluminium and chromium. Readers requiring more detail of these are referred to the review by Wilson (1975).

In ion implantation processes, ions with very high energies, of the order of MeV, are fired into the component surface and, as a result of their high energies, penetrate the surface up to distances of $\simeq 1\ \mu$m. Very high concentrations of the implanted ions can be created in this way and the resulting surfaces often have excellent wear resistance (Dearnaley and Grant, 1985). One beneficial, and as yet unexplained, effect is that the improved wear resistance often persists up to depths well below the original depth of implantation.

### Application of Surface Coatings

Surface coatings can be applied over a range of thicknesses from several millimetres to a few micrometres. At the thicker end of this range, layers of wear-resistant metal can be deposited by welding techniques, and this process can be used to reclaim very badly worn components such as the teeth of excavators. Popular materials for this type of application are the cobalt-based superalloys described in Section 4.1.

Coatings of the order of 1 mm thick can be deposited by flame spraying, which is usually used for metallic coatings; plasma spraying, which can be used for a wide range of metals and ceramics; and spray-fuse coating, where a metal is initially sprayed as a powder and subsequently fused, to form a fully dense and highly adherent coating.

Coatings of the order of 0.1 mm thick can be deposited by electroplating, electroless deposition, plasma spraying or detonation-gun plating. Electroplating is used to deposit hard and abrasion-resistant chromium plating, and also to deposit the lead alloy overlays on bearing alloys. Electroless deposition can be used to deposit a range of hard and wear-resistant nickel-based coatings. Both plasma spraying and detonation-gun plating can be used to deposit a wide range of ceramic and cermet coatings, which are finding extensive use as wear- and corrosion-resistant materials in hostile environments such as nuclear reactors. Both types of coating are slightly porous, the plasma-sprayed coatings being rather worse in this respect, and this can allow some aggressive corrosive species to penetrate the coating and cause it to separate from its substrate.

Finally, coatings only a few micrometres thick are being increasingly deposited by physical vapour deposition (Teer and Arnell, 1985) and chemical vapour deposition (Hitchman, 1985). Both these techniques can allow the deposition of fully dense and highly adherent coatings of a wide range of compositions, including metastable compositions which cannot be formed by any other means. Such coatings are already being spectacularly successful in a variety of applications, which include: the coating of cutting tools with titanium nitride to give a fivefold increase in life; coating of the races of satellite bearings to give a coating $\simeq 0.5$ $\mu$m thick which provides soft metal film lubrication for the full life of the satellite; and coating of other satellite bearings with highly adherent films of molybdenum disulphide.

## 4.8 PROBLEMS

No problems have been set for this chapter. This is because exhaustive design procedures for dry rubbing bearings have been set out recently in ESDU Item No. 87007: *Design and Material Selection for Dry Rubbing Bearings*. For commercial reasons neither the data nor the procedures can be reproduced here, and under these circumstances it is not possible to set meaningful design problems. However, the reader is strongly recommended to study the ESDU unit and to work through the example design problems therein.

# Chapter 5
# Fluid-lubricated Thrust Bearings

## 5.1 INTRODUCTION

So far in this book the contact of bodies in relative transverse or sliding motion has been considered for the important case of sensibly dry contact. In such situations, which are common in many everyday applications, there is by definition intimate contact between the sliding, or rubbing, pairs of surfaces, and the consequences of this physical fact have been discussed at length in the previous chapters.

However, sliding surfaces may be completely separated by the creation of relatively thin films of fluid which are capable of transmitting the normal loads between the surfaces while allowing sliding to occur in the transverse direction. The study of such films is based on hydrodynamic theory as applied to the analysis of the viscous flow of fluids in gaps or clearances which are relatively small in comparison with the other dimensions of the components involved. The application of the results of these studies leads to the design and development of *hydrodynamic*, or self-acting, fluid film bearings.

In this chapter much simplified analyses will be made of relatively simple geometric configurations in order to gain an understanding of the principles involved, but the results of much more complicated solutions will be presented and used in the design of hydrodynamic thrust bearings. This approach will be extended in the next chapter to show how the design of hydrodynamic journal bearings may also be approached.

Fluid films may also be created between sliding surfaces by supplying fluid from an external supply at a relatively high pressure. Such externally pressurised, or *hydrostatic*, bearings are usually used in special circumstances and have been extensively studied (Opitz, 1967; Halling, 1975a). This class of bearing will not be dealt with in this book.

## 5.2 FLUID PROPERTIES

### 5.2.1 Dynamic Viscosity

The physical property of state known as dynamic viscosity was recognised by Newton, some 300 years ago, as an important characteristic of fluids (Massey, 1989). Dynamic viscosity may be defined, as far as the requirements of this chapter are concerned, by reference to Figure 5.1, where a plate of area $A$ is propelled with a force $F$ at a velocity $U$ so that a film of fluid of uniform thickness $h$ exists between the plate and a plane stationary surface. The axes $x$ and $z$ are introduced as shown, to describe the geometry of the problem.

The shear stress, $\tau$, which must act on the underside of the plate is given by $\tau = F/A$. The velocity gradient, $\partial u/\partial z$, is assumed to be constant throughout the thickness of the fluid film and is given by $\partial u/\partial z = U/h$. Newton proposed that for an ideal fluid the shear stress, $\tau$, is proportional to the velocity gradient $\partial u/\partial z$. Thus,

$$T \propto \frac{\partial u}{\partial z} \qquad \text{or} \qquad \tau = \eta \frac{\partial u}{\partial z} \qquad (5.1)$$

where the constant of proportionality, $\eta$, is known as the coefficient of dynamic viscosity. The dimensions of this coefficient may be found as $\eta = \tau/(\partial u/\partial z)$ and it follows that $\tau/(\partial u/\partial z)$ has dimensions of

$$\frac{(F/L^2)}{(L/T)/L} = \frac{FT}{L^2}$$

In the SI system the coefficient of dynamic viscosity has units of (newton·second)/(metre)$^2$, but as a newton is defined as (kilogram) (metre/second$^2$), the units of dynamic viscosity may also be given in dimensions of $(ML/T^2)T/L^2 = M/LT$ or (kilogram)/(metre·second). For example, water at room temperature and atmospheric pressure has a dynamic viscosity of about $10^{-3}$ Ns/m$^2$ or $10^{-3}$ kg/m·s.

Another common and important unit of dynamic viscosity is the poise, which has units of (dyne·second)/(centimetre)$^2$ or (gram)/(centimetre·second) in the CGS system of units, but a more useful unit is the centipoise (cP). The relationships between these units is as follows:

$$\text{dynamic viscosity (kg/m·s)} = \tfrac{1}{10} \text{ dynamic viscosity (P)}$$
$$= \tfrac{1}{1000} \text{ dynamic viscosity (cP)} \qquad (5.2)$$

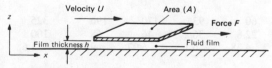

**Figure 5.1**

## 5.2.2 Kinematic Viscosity

The kinematic viscosity of a fluid is simply defined as the ratio between its dynamic viscosity, $\eta$, and its density, $\rho$. So the kinematic viscosity, $v$, is simply $v = \eta/\rho$ and its dimensions in the SI system are $(\text{metre})^2/(\text{second})$. The corresponding unit in the CGS system has units of $(\text{centimetre})^2/(\text{second})$ and is known as the stoke, although a more useful unit is the centistoke (cSt). This latter unit is very important, because most of the internationally

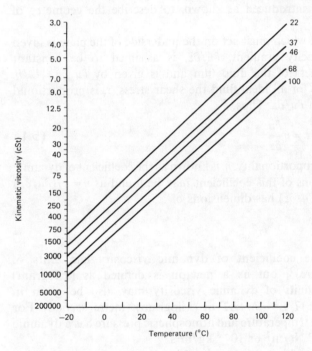

**Figure 5.2** Typical characteristics of Shell Tellus oils:

| | | Shell Tellus oil | | | | |
|---|---|---|---|---|---|---|
| | | 22 | 37 | 46 | 68 | 100 |
| Viscosity, kinematic, cSt, IP 71 | at 0°C | 188 | 340 | 535 | 960 | 1650 |
| | at 20°C | 60 | 93 | 170 | 196 | 325 |
| | at 40°C | 22.8 | 36.2 | 46 | 68 | 100 |
| | at 100°C | 4.38 | 6.1 | 6.97 | 9.05 | 11.6 |
| Density at 15°C, kg/l, IP 160 | | 0.866 | 0.875 | 0.878 | 0.880 | 0.885 |

recognised scientific instruments which are used to measure viscosity are calibrated in centistokes (cSt).

However, it should be noted that there is also a range of instruments, known as efflux viscometers, which measure viscosity by simply giving the time required for a given volume of fluid to flow through an orifice of known characteristics. Thus, the viscosity of engine oils, for example, is often quoted in units of 'seconds' as measured by a No. 1 Redwood viscometer. It is usually possible to convert such measurements in 'seconds' into the more technically useful units of centistokes, for example (Massey, 1989).

## 5.2.3   Variation of Viscosity with Temperature

A major difficulty in the design of hydrodynamic bearings is the rapidity with which the viscosity of mineral oils, with and without additives, changes with temperature. This feature is important, as mineral oils are used extensively to lubricate such bearings. The caption and graphs of Figure 5.2 show this variation for a family of commercially available oils, the Shell Tellus oils. It should be noted that the scales of the graphs have been distorted, to produce straight-line characteristics, as these are of practical use. A more revealing presentation of the viscosity–temperature characteristics for two of this family of oils is given in Figure 5.3.

The variation of density with temperature of mineral oils is negligible in comparison with the variation of viscosity, and the values of density given in the caption of Figure 5.2 may be taken to be constant over normal ranges of temperature. Similarly, the changes in pressure experienced by oils in hydrodynamic bearings are normally not large enough to affect their viscosity, but this is not the case in highly loaded non-conforming contacts, as discussed in Chapter 7.

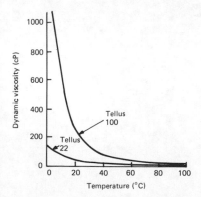

**Figure 5.3**   Viscosity–temperature characteristics on linear scales

## 5.3 LAMINAR FLOW THEORY

### 5.3.1 Flow through a Parallel Gap

Consider the flow of a viscous fluid (Massey, 1989) through a wide parallel rectangular gap of relatively small, but uniform, clearance $h$ under the action of a pressure drop from $p_1$ to $p_2$ over a length $L$ and the action of a moving surface which has a velocity $U_c$, as shown in Figure 5.4. The axes $x$, $y$ and $z$ are set up as before and it is assumed that the edges at $y = 0$ and $y = B$, corresponding to the width of the gap, are sealed so that the flow of fluid proceeds only in the direction of the $x$ axis.

Consider the equilibrium of an element of fluid of length $\delta x$ at a distance $x$ from the origin. Let this element have a thickness $\delta z$ and be positioned at a height of $z$ above the lower surface which moves to the right with a velocity $U_c$. The local pressure at $x = 0$ is $p_1$, and at $x = L$, $p_2$.

The element of fluid of Figure 5.4 is shown enlarged in Figure 5.5, together with the pressures and shear stresses which are assumed to act on it. Pressure $p$ is assumed to increase with $x$ and the shear stress $\tau$ with $z$.

In the steady state the sum of the forces acting on the element is zero. In each case the local pressure, or shear stress, may be multiplied by the appropriate area, to give the corresponding force. The sum of the forces acting to the right on the element is thus:

$$p(B\delta z) + \left(\tau + \frac{\partial \tau}{\partial z} \cdot \delta z\right)(B\delta x) - \left(p + \frac{\partial p}{\partial x} \cdot \delta x\right)(B\delta z) - \tau(B\delta x) = 0$$

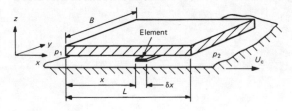

**Figure 5.4**

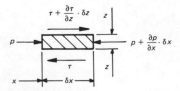

**Figure 5.5**

or

$$\frac{\partial \tau}{\partial z} = \frac{\partial p}{\partial x} \tag{5.3}$$

This is the equation of equilibrium for the element of fluid. Development of the solution now proceeds on the basis that the flow throughout the gap is laminar, or viscous, so that the relationship between the shear stress and velocity gradient given in Equation (5.1) may be substituted into Equation (5.3) above, to yield the following expression:

$$\eta \frac{\partial^2 u}{\partial z^2} = \frac{\partial p}{\partial x}$$

As $\partial p/\partial x$ is not a function of $z$, this equation may be integrated twice, to obtain:

$$u = \left(\frac{1}{\eta} \frac{\partial p}{\partial x}\right) \frac{z^2}{2} + C_1 z + C_2$$

where $C_1$ and $C_2$ are constants to be determined. Now, when $z = 0$, we know that $u = U_c$, and when $z = h$, $u = 0$.

Thus

$$C_1 = -\left(\frac{1}{\eta} \frac{\partial p}{\partial x}\right) \frac{h}{2} - \frac{U_c}{h}$$

and $C_2 = U_c$. These values for the constants may be back-substituted into the above expression, to give the following solution for the velocity of the fluid in the gap:

$$u = \frac{1}{2\eta}\left(-\frac{\partial p}{\partial x}\right)(hz - z^2) + U_c\left(1 - \frac{z}{h}\right) \tag{5.4}$$

It is clear that the pressure gradient must be negative if the flow is to proceed to the right and that the total velocity at any value of $z$ is given by the sum of the pressure-induced, or Poiseuille, velocity and the shear velocity induced by the movement of the lower surface.

The total volumetric flow rate of fluid through the gap is given as $Q = \int_0^h uB\delta z$. The expression for $u$, a function of $z$, may be substituted from Equation (5.4) into this expression to give

$$Q = B \int_0^h \left\{ \frac{1}{2\eta}\left(-\frac{\partial p}{\partial x}\right)(hz - z^2) + U_c\left(1 - \frac{z}{h}\right)\right\} dz$$

or

$$Q = \frac{Bh^3}{12\eta}\left(-\frac{\partial p}{\partial x}\right) + \frac{Bh}{2} U_c \tag{5.5}$$

In the case of the geometry of Figure 5.4, it is clear that the volumetric flow rate given by Equation (5.5) does not vary with $x$, so that

$$\frac{\partial p}{\partial x} = -12\eta\left(\frac{Q}{B} - \frac{h}{2}U_c\right)$$

Integration with respect to $x$ gives

$$p = 12\eta\left(\frac{Q}{B} - \frac{h}{2}\cdot U_c\right)x + C_3$$

Now, when $x = 0$, $p = p_1$, so $C_3 = p_1$, and when $x = L$, $p = p_2$; thus,

$$p_2 = -12\eta\left(\frac{Q}{B} - \frac{h}{2}U_c\right)L + p_1$$

In this case the pressure gradient, $\partial p/\partial x$, is a constant equal to $(p_2 - p_1)/L$; note that this is negative, so that for this special case Equation (5.5) becomes

$$Q = \frac{Bh^3}{12\eta}\left(\frac{p_1 - p_2}{L}\right) + \frac{Bh}{2}U_c \tag{5.6}$$

### 5.3.2 Example of Flow Rate Calculation

Tellus 37 oil flows at a mean temperature of 40°C through a rectangular channel or gap of uniform thickness 0.1 mm, width 15 mm and length 200 mm under a pressure drop of 4 $N/mm^2$. What is the flow rate of fluid?

From Figure 5.2 the kinematic viscosity of the oil is 36.2 cSt and its density 0.875 $g/cm^3$. Its dynamic viscosity in SI units is, from Equation (5.2),

$$\frac{36.2 \times 0.875}{1000} = 0.0317 \text{ kg/m}\cdot\text{s}$$

In this case $U_c = 0$ and the flow rate follows from Equation (5.6) as

$$Q = \frac{(15 \times 10^{-3})(0.1 \times 10^{-3})^3}{12 \times 0.0317}\left\{\frac{4 \times 10^6 - 0}{200 \times 10^{-3}}\right\} = 0.807 \times 10^{-6} \text{ m}^3/\text{s}$$

$$= 0.048 \text{ l/min}$$

## 5.4 HYDRODYNAMIC THRUST BEARINGS

### 5.4.1 Plane-inclined Slider Bearings – Reynolds's Equation

This elementary geometry is shown in Figure 5.6 and it is assumed that the width of the bearing, $B$, is much greater than its length, $L$. This restriction has the effect of ensuring that most of the flow through the gap between the

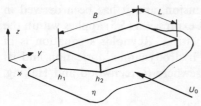

**Figure 5.6**

bearing and the moving surface is in the direction of the $x$ axis. It will only be near the ends of the bearing, where $y = 0$ and $B$, that there will be significant flow in the direction of the $y$ axis. This restriction is discussed further in sub-section 5.6.1.

There is no change of gap or film thickness in the direction of the $y$ axis, but an uniform inclination of the undersurface of the bearing is assumed to exist in the direction of the $x$ axis, as shown in the cross-section of Figure 5.7. The gap is shown to vary uniformly from a minimum of $h_1$ to a maximum of $h_2$ over the length, $L$, of the bearing. The *stationary* origin is taken to be at the intersection of the produced upper stationary and the lower moving surfaces of the bearing.

Now let there be a local pressure $p$, a pressure gradient $\partial p/\partial x$ and a film thickness $h$ at the section of the bearing located by the ordinate $x$. If the fluid is incompressible, then the volumetric flow rate of fluid through the bearing must be a constant. This is the continuity of flow statement, which means that as much fluid flows into the bearing at one end as flows out at the other, so the flow rate, $Q$, is a constant and independent of $x$; thus, $\partial Q/\partial x = 0$. Substituting for $Q$ from Equation (5.5) in this expression gives

$$\frac{\partial}{\partial x}\left\{\frac{h^3}{12\eta}\left(-\frac{\partial p}{\partial x}\right)+\frac{h}{2}U_c\right\}=0$$

This expression may be put in the following form, as $U_c$ is not a function of $x$:

$$\frac{\partial}{\partial x}\left\{\frac{h^3}{\eta}\frac{\partial p}{\partial x}\right\}=6U_c\frac{\partial h}{\partial x} \tag{5.7}$$

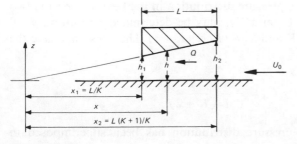

**Figure 5.7**

This is Reynolds's equation in one dimension, and it has been derived in this very simple manner by assuming that expression (5.5) applies within the film even though its thickness changes with $x$. Reynolds's equation is of fundamental importance in the study of hydrodynamic lubrication and its derivation is discussed in more detail elsewhere (Greenwood and Halling, 1971; Cameron, 1981).

In the case of Figure 5.7 it is clear that

$$\frac{\partial h}{\partial x} = \frac{h_2 - h_1}{L} = \frac{h_1}{L}\left(\frac{h_2}{h_1} - 1\right) = \frac{h_1}{L} \cdot K$$

where $K = (h_2/h_1) - 1$. Study of the geometry also shows that $h = h_1$ when $x_1 = L/K$ and $h = h_2$ when $x_2 = (L/K)(K + 1)$.

Putting $h = (\partial h/\partial x)x = (h_1 K/L)x$ and $U_c = -U_0$ in Reynolds's equation (5.7) gives the following equation:

$$\frac{\partial}{\partial x}\left\{\left(\frac{h_1 x}{L}\right)^3 \frac{x^3}{\eta}\frac{\partial p}{\partial x}\right\} = -6U_0\left(\frac{h_1 K}{L}\right)$$

or

$$\left\{\left(\frac{h_1 K}{L}\right)^2 \frac{1}{6\eta U_0}\right\}\frac{\partial}{\partial x}\left(x^3 \frac{\partial p}{\partial x}\right) = -1$$

Note that the terms in the first parentheses may be assumed to be constant for a given bearing if the variation of viscosity with temperature is ignored at this stage. The above expression may now be integrated twice with respect to $x$, to obtain the following expressions:

$$\left\{\frac{1}{6\eta U_0}\left(\frac{h_1}{L}\right)^2\right\}K^2\frac{\partial p}{\partial x} = -\frac{1}{x^2} + \frac{C_3}{x^3} \tag{5.8}$$

$$\left\{\frac{1}{6\eta U_0}\left(\frac{h_1}{L}\right)^2\right\}K^2 p = \frac{1}{x} - \frac{C_3}{2x^2} + C_4 \tag{5.9}$$

This is the solution for the pressure distribution in the bearing, and the two constants of integration, $C_3$ and $C_4$, may be determined by putting $p = 0$ (atmospheric) at $x_1 = L/K$ and $x_2 = (L/K)(K + 1)$. The constants have the following values:

$$C_3 = 2\frac{L}{K}\left(\frac{K + 1}{K + 2}\right); \qquad C_4 = -\frac{K}{L}\left(\frac{1}{K + 2}\right)$$

A typical shape of this pressure distribution has been superimposed in Figure 5.7.

Of interest is the value of the maximum pressure which occurs at $x_m$ where $\partial p/\partial x = 0$. It follows from Equation (5.8) that:

$$x_m = \left\{ \frac{2}{K} \left( \frac{K+1}{K+2} \right) \right\} L$$

and that the corresponding value of the film thickness is $h_m$, where

$$h_m = \left( \frac{h_1 K}{L} \right) x_m = \left\{ 2 \left( \frac{K+1}{K+2} \right) \right\} h_1$$

The maximum value of the pressure in the film at $x_m$ is found from Equation (5.9) as

$$p_m = \left( \frac{\eta U_0 L}{h_1^2} \right) \left( \frac{3K}{2(K+1)(K+2)} \right)$$

Now the total load carried by the pad, $W$, is given by the following integral:

$$W = \int_{x_1}^{x_2} pB \, dx = \left\{ 6\eta U_0 \left( \frac{L}{h_1} \right)^2 \right\} \frac{B}{K^2} \int_{L/K}^{(L/K)(K+1)} \left( \frac{1}{x} - \frac{C_3}{2x^2} + C_4 \right) dx$$

By substituting for the constants $C_3$ and $C_4$ and completing the above integral, it may be shown that

$$W = \left( \eta U_0 B \left( \frac{L}{h_1} \right)^2 \right) \cdot \left\{ \frac{6}{K^2} \left( \log_e(K+1) - \frac{2K}{K+2} \right) \right\}$$

The rate of flow of fluid through the bearing may be found conveniently at $x_m$, as there is no pressure-induced flow here, because the pressure gradient is zero. The velocity- or shear-induced flow is then given by the second term of Equation (5.4) as $Q = (U_0/2)Bh_m$ and putting

$$h_m = 2 \left( \frac{K+1}{K+2} \right) h_1$$

gives the flow rate as:

$$Q = (Bh_1 U_0) \left( \frac{K+1}{K+2} \right)$$

Last, the lower sliding member experiences a shear force due to the shear stress $\tau$ which is distributed over its surface as it passes under the inclined pad. Remembering that the shear stress is given by $\tau = \eta \, \partial u/\partial x$ and that $u$ is given by expression (5.3), it follows that, at $z = 0$,

$$\tau = \eta \frac{\partial u}{\partial x} = -\frac{h}{2} \frac{\partial p}{\partial x} + \eta \frac{U_0}{h}$$

The total shear force on the sliding member is given by

$$F = \int_{x_1}^{x_2} \tau B \, dx = B \int_{L/K}^{(L/K)(K+1)} \left( -\frac{h}{2}\frac{\partial p}{\partial x} + \eta \frac{U_0}{h} \right) dx$$

Integrating the first term by parts and the second by putting $h = (h_1 K/L)x$ gives

$$F = W\frac{h_1}{L}\left(\frac{K}{2}\right) + \left(\frac{\eta U_0 B}{h_1}\right)\left(\frac{\log_e(K+1)}{K}\right)$$

Substituting for the previously determined expression for $W$,

$$F = \left(\frac{\eta U_0 BL}{h_1}\right)\left\{\frac{4}{K}\log_e(K+1) - \frac{6}{K+2}\right\}$$

## 5.4.2 Summary of Results – Dimensionless Coefficients

The detailed analysis presented in the previous section has shown that, within the inherent assumptions, it has been possible to derive expressions for four of the main parameters which would be of interest in the design of a bearing of the geometry under discussion. These parameters are the maximum pressure in the film, the volumetric flow rate of fluid required to operate the bearing, the load capacity of the bearing and the shear force necessary to move the sliding member under the bearing.

It will be noticed that the fluid film converges in the direction of sliding and that all of the four parameters listed above are proportional to the sliding velocity. Thus, the pressures under the bearing are developed by the action of the sliding member dragging fluid through the converging gap – i.e. the bearing is self-acting.

The four expressions are reproduced in a convenient form in expressions (5.10):

$$\left.\begin{aligned}
p_m &= \left(\frac{\eta U_0 L}{h_1^2}\right)\left(\frac{3K}{2(K+1)(K+2)}\right) \\[2mm]
Q &= (Bh_1 U_0)\left(\frac{K+1}{K+2}\right) \\[2mm]
W &= \left\{\eta U_0 B\left(\frac{L}{h_1}\right)^2\right\}\left[\frac{6}{K^2}\left\{\log_e(K+1) - \frac{2K}{K+2}\right\}\right] \\[2mm]
F &= \left(\frac{\eta U_0 BL}{h_1}\right)\left\{\frac{4}{K}\log_e(K+1) - \frac{6}{K+2}\right\}
\end{aligned}\right\} \qquad (5.10)$$

Each of the four practically important parameters is given as the product of two separate terms. In each case the first set of enclosures contains all *dimensional* information for the particular bearing, while the second term contains information which depends only on the shape or geometry of the bearing in *dimensionless* form.

In this simple case the geometry of the bearing (Figure 5.6) is described by a single *dimensionless* variable, $K$, which is a function of the ratio between the entry and exit film thicknesses, $h_2$ and $h_1$. Therefore, the following four *dimensionless* parameters or coefficients are defined and used to simplify the presentation of the corresponding expressions of (5.10):

$$\left.\begin{aligned}
\bar{p}_m &= \left\{ \frac{3K}{2(K+1)(K+2)} \right\} \\
\bar{Q} &= \left( \frac{K+1}{K+2} \right) \\
\bar{W} &= \left[ \frac{6}{K^2} \left\{ \log_e(K+1) - \frac{2K}{K+2} \right\} \right] \\
\bar{F} &= \left\{ \frac{4}{K} \log_e(K+1) - \frac{6}{K+2} \right\}
\end{aligned}\right\} \quad (5.11)$$

The expressions of (5.10) then become:

$$\left.\begin{aligned}
p_m &= \bar{p}_m \left( \frac{\eta U_0 L}{h_1^2} \right) \\
Q &= \bar{Q}(Bh_1 U_0) \\
W &= \bar{W} \left\{ \eta U_0 B \left( \frac{L}{h_1} \right)^2 \right\} \\
F &= \bar{F} \left( \frac{\eta U_0 BL}{h_1} \right)
\end{aligned}\right\} \quad (5.12)$$

The four important parameters for a wide hydrodynamic thrust bearing under discussion are therefore given by the straightforward expressions (5.12), where the *dimensional* details of any particular bearing are given by the contents of the enclosures, while the shape or geometry of the bearing is reflected in the values of the four *dimensionless* coefficients $\bar{p}_m$, $\bar{Q}$, $\bar{W}$ and $\bar{F}$, which multiply the sets of enclosed terms.

In this case the values of the *dimensionless* coefficients $\bar{p}_m$, $\bar{Q}$, $\bar{W}$ and $\bar{F}$ depend only on one geometric parameter, $K$, which itself is dependent on the ratio $h_2/h_1$. The values of these four coefficients may be evaluated from expressions (5.11) and tabulated for a useful range of $K$, as shown in Figure 5.8, but it is more usual to express these results in the graphical form of Figure 5.9. It may be shown from the third expression of (5.11) that the maximum value of $\bar{W}$ occurs when $K = 1.1889$.

### 5.4.3 Example of Thrust-bearing Calculations

A thrust bearing of width 200 mm and length 50 mm operates with a minimum film thickness of 0.1 mm and a sliding velocity of 3 m/s. If the effective viscosity of the lubricating oil is assumed for simplicity to be 40 cP and the film thickness ratio is adjusted to produce the maximum load capacity, determine the operating characteristics of the bearing.

As the maximum load capacity is produced when $K = 1.1889$, we get $h_2/h_1 = K + 1 = 2.1889$, and as $h_1 = 0.1$ mm, $h_2$ is 2.1889 mm. The value of $\bar{p}_m$, $\bar{Q}$, $\bar{W}$ and $\bar{F}$ corresponding to this value of $K$ may therefore be evaluated from Equations (5.11) or read from Table 5.1 and Figure 5.8.

The value of viscosity to be used in the calculations may be found in SI units from expression (5.2) as $\frac{40}{1000} = 0.04$ kg/m·s.

*Maximum pressure*

From Figure 5.8, $\bar{p}_m = 0.2555$ and expressions (5.12) give

$$p_m = 0.2555 \left\{ \frac{0.04 \times 3(50 \times 10^{-3})}{(0.1 \times 10^{-3})^2} \right\} = 153 \times 10^3 \, \text{N/m}^2 = 0.153 \, \text{N/mm}^2$$

*Flow rate*

From Figure 5.8, $\bar{Q} = 0.6804$ and expressions (5.12) give

$$Q = 0.6804 \{ (200 \times 10^{-3})(0.1 \times 10^{-3})3 \}$$

$$= 41.184 \times 10^{-6} \, \text{m}^3/\text{s} = 2.47 \, \text{l/min}$$

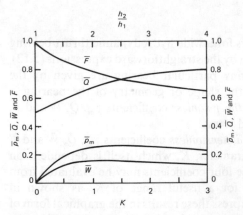

**Figure 5.8** Variation of dimensionless coefficients, $\bar{p}_m$, $\bar{Q}$, $\bar{W}$ and $\bar{F}$ for a wide hydrodynamic thrust bearing with film thickness ratio $h_2/h_1$

**Table 5.1** Variation of dimensionless coefficients $\bar{p}_m$, $\bar{Q}$, $\bar{W}$ and $\bar{F}$ for a wide hydrodynamic thrust bearing with film thickness ratio $h_2/h_1$

| $\dfrac{h_2}{h_1}$ | 1.25 | 1.5 | 1.75 | 2.0 | 2.1889 | 2.25 | 2.5 | 2.75 | 3.0 | 3.25 | 3.5 | 3.75 | 4.0 |
|---|---|---|---|---|---|---|---|---|---|---|---|---|---|
| $K$ | 0.25 | 0.5 | 0.75 | 1.0 | 1.1889 | 1.25 | 1.5 | 1.75 | 2.0 | 2.25 | 2.5 | 2.75 | 3.0 |
| $\bar{p}_m$ | 0.1333 | 0.2000 | 0.2338 | 0.2500 | 0.2555 | 0.2564 | 0.2571 | 0.2545 | 0.2500 | 0.2443 | 0.2381 | 0.2316 | 0.2250 |
| $\bar{Q}$ | 0.5556 | 0.6000 | 0.6364 | 0.6667 | 0.6804 | 0.6923 | 0.7143 | 0.7333 | 0.7500 | 0.7647 | 0.7778 | 0.7895 | 0.8000 |
| $\bar{W}$ | 0.0884 | 0.1312 | 0.1511 | 0.1589 | 0.1602 | 0.1601 | 0.1577 | 0.1533 | 0.1479 | 0.1420 | 0.1360 | 0.1300 | 0.1242 |
| $\bar{F}$ | 0.9036 | 0.8437 | 0.8028 | 0.7726 | 0.7542 | 0.7488 | 0.7292 | 0.7122 | 0.6972 | 0.6836 | 0.6711 | 0.6594 | 0.6494 |

*Load capacity*

Again Figure 5.8 gives $\bar{W} = 0.1602$ and, using expressions (5.12),

$$W = 0.1602 \left\{ 0.04 \times 3(200 \times 10^{-3}) \left( \frac{50 \times 10^{-3}}{0.1 \times 10^{-3}} \right)^2 \right\} = 960 \text{ N}$$

*Shear force*

Figure 5.8 gives $\bar{F} = 0.7542$ and, using expressions (5.12),

$$F = 0.7542 \left\{ \frac{0.04 \times 3(200 \times 10^{-3})(50 \times 10^{-3})}{(0.1 \times 10^{-3})} \right\} = 9.05 \text{ N}$$

### 5.4.4   Further Characteristics of Plane-inclined Slider Bearings

The coefficient of 'friction' of a bearing is sometimes defined as $\mu$, the ratio between the shear force and load capacity – i.e. $\mu = F/W$. In the previous example we get $\mu = F/W = 9.05/960 = 0.0094$. This low value is typical of self-acting hydrodynamic bearings.

Of great practical interest is the power required to drive the sliding member under the bearing and this is simply the product of the shear force and the sliding velocity – i.e. $P = FU_0$. In the previous example the power is $P = FU_0 = 9.05 \times 3 = 27.15 \text{ W}$.

Because of the sensitivity of the viscosity of mineral oils to temperature, it is again of great practical importance to assess the temperature rise of the oil as it passes through the bearing. The power consumed by the bearing appears as heat which causes a rise in the temperature of oil. It will be assumed in the first instance that all of the oil which flows through the bearing is equally effective in absorbing the heat which is generated in the bearing. It will also be assumed that no heat is conducted through the metallic components of the bearing, so that all of the heat generated is absorbed by the oil.

A simple power balance may then be written as

$$P = FU_0 = Q\rho C_p \Delta T$$

where $\rho$ is the density of the oil, $C_p$ its specific heat and $\Delta T$ the temperature rise of the oil. Most mineral oils have a specific heat of about 1880 Nm/kg·°C. Hence,

$$\Delta T = \frac{P}{Q\rho C_p} = \frac{27.5}{(41.184 \times 10^{-6})(880)(1880)} = 0.4°C$$

Figure 5.9 has been produced to summarise the discussion of this bearing.

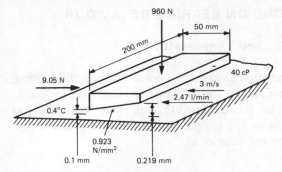

**Figure 5.9**

## 5.4.5 Further Dimensionless Coefficients for a Plane-inclined Slider Bearing

Other useful coefficients may be derived from those which have already been presented. For example, the power balance may be modified by recognising that only a proportion, $k$, of the power is convected out of the bearing by the oil. Then

$$kP = kFU_0 = Q\rho C_p \Delta T$$

so,

$$k\bar{F}\left(\frac{\eta U_0 BL}{h_1}\right)U_0 = \bar{Q}(Bh_1 U_0)\rho C_p \Delta T$$

Putting $\bar{T} = \bar{F}/\bar{Q}$,

$$\Delta T = k\bar{T}\left(\frac{\eta U_0 L}{\rho C_p h_1^2}\right) \tag{5.13}$$

Thus, the temperature rise is given directly as the product of a dimensional group, in parentheses, and a dimensionless coefficient which is dependent on the geometry of the bearing. In the previous example, $\bar{T} = \bar{F}/\bar{Q} = 0.7542/0.6804 = 1.0988$, and taking $k = 1$, although a more typical value is about 0.8, we get from expression (5.13)

$$\Delta T = k\bar{T}\left(\frac{\eta U_0 L}{\rho C_p h_1^2}\right) = 1 \times 1.0988\left\{\frac{0.4 \times 3(50 \times 10^{-3})}{880 \times 1880(0.1 \times 10^{-3})^2}\right\}$$

$$= 0.4\,°\mathrm{C}, \text{ as before}$$

Values of $\bar{T}$ for the bearing in question have been added to Figure 5.9.

## 5.5 EFFECTS OF HEATING ON BEARING BEHAVIOUR

### 5.5.1 Effective Viscosity and Temperature

As the temperature of the lubricant increases as it passes through the bearing, it is obviously necessary to compensate for the inevitable decrease of viscosity which results in all mineral oils. It is assumed that oil enters the bearing at an inlet temperature $T_i$ and leaves at a temperature $T_o$. The temperature rise, $\Delta T$, is thus found from the power balance as

$$\Delta T = T_o - T_i = k\bar{T}\left\{\frac{\eta U_0 L}{\rho C_p h_1^2}\right\}$$

The effective temperature, $T_e$, of the lubricant is conventionally taken as the mean of the inlet and outlet temperatures:

$$T_e = \frac{T_o + T_i}{2} = T_i + \frac{\Delta T}{2}$$

It is clearly necessary to relate the effective temperature, $T_e$, of the oil to the operating characteristics of the bearing, and it is normal to take the effective viscosity of the oil as that at the effective temperature, $T_e$. It is assumed that for the case of a wide plane-inclined slider bearing all of the oil which enters the bearing travels through the whole length, $L$, of the bearing and is therefore all equally effective in cooling the bearing, as comparatively little oil escapes from the bearing in the sideways direction.

The following examples illustrate the methods which may be used to include the viscosity–temperature characteristic of an oil in the assessment of a bearing's behaviour.

### 5.5.2 Example of Thrust-bearing Calculations Including Effects of Temperature

A plane-inclined pad thrust bearing is 160 mm wide and 40 mm long in the direction of sliding. The sliding velocity is 6 m/s and the bearing is supported so that the inlet film thickness is *always* three times the outlet film thickness. A load of 18 kN is carried by the bearing and Tellus 37 oil is supplied at an inlet temperature of 40°C. Assuming that 80% of the heat generated in the bearing is carried away by the oil, describe the operating characteristics of the bearing.

From Figure 5.2 the required table and graph may be derived (Figure 5.10).

As $h_2/h_1 = 3$, we get $K = 2$, $\bar{p}_m = 0.25$, $\bar{Q} = 0.75$, $\bar{W} = 0.1479$, $\bar{F} = 0.6972$ and $\bar{T} = 0.9296$ from Figure 5.8.

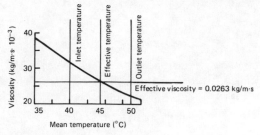

**Figure 5.10**

From expressions (5.12) and (5.13) we may write

$$W = \bar{W}\left\{\eta U_0 B\left\{\frac{L}{h_1}\right\}^2\right\}$$

and

$$\Delta T = k\bar{T}\left(\frac{\eta U_0 L}{\rho C_p h_1^2}\right)$$

Inserting all of the known information yields

$$18\,000 = 0.1479\left\{\eta \times 6 \times 0.16\left(\frac{0.04}{h_1}\right)^2\right\}$$

and

$$\Delta T = 0.8 \times 0.9296\left(\frac{\eta \times 6 \times 0.04}{875 \times 1880 h_1^2}\right)$$

which reduce to the following expressions:

$$18\,000 = 2.272 \times 10^{-4}\left(\frac{\eta}{h_1^2}\right)$$

and

$$\Delta T = 10.85 \times 10^{-6}\left(\frac{\eta}{h_1^2}\right)$$

Eliminating $\eta/h_1^2$ from these two expressions gives

$$\Delta T = 10.85 \times 10^{-6} \times \frac{18\,000}{2.272 \times 10^{-4}} = 8.59°C$$

So the outlet temperature is $40 + 8.59 = 48.59°C$ and the effective temperature is given as

$$T_e = T_i + \frac{\Delta T}{2} = 40 + \frac{8.59}{2} = 44.3°C$$

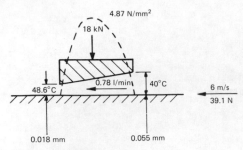

**Figure 5.11**

Reference to Figure 5.2 gives the effective viscosity at the effective temperature of 44.3°C as 0.0263 kg/m·s. The corresponding value of the film thickness $h_1$ then follows from the above as

$$h_1 = \left(10.85 \times 10^{-6}\, \frac{\eta}{\Delta T}\right)^{\frac{1}{2}} = \left(\frac{10.85 \times 10^{-6} \times 0.0263}{8.59}\right)^{\frac{1}{2}}$$

$$h_1 = 1.82 \times 10^{-5}\, \text{m} = 0.018\, \text{mm}$$

So,

$$h_2 = 3h_1 = 3(0.018) = 0.054\, \text{mm}$$

The important operating characteristics then follow from expressions (5.12) as

$$p_m = 0.25 \left\{\frac{0.0263 \times 6 \times 0.04}{(0.018 \times 10^{-3})^2}\right\} = 4.87 \times 10^6\, \text{N/m}^2$$

$$Q = 0.75(0.16 \times 0.018 \times 10^{-3} \times 6) = 12.96 \times 10^{-6}\, \text{m}^3/\text{s}$$

$$F = 0.6972 \left(\frac{0.0263 \times 6 \times 0.16 \times 0.04}{0.018 \times 10^{-3}}\right) = 39.1\, \text{N}$$

The operating characteristics of the bearing are shown in Figure 5.11.

### 5.5.3 Bearings with a Known Taper

In the previous example it was assumed that the dimensionless parameters of the bearing were known, as it was stated that the manner in which the bearing was supported ensured that the ratio between the inlet and outlet film thicknesses was always equal to 3. This restraint may be imposed on a plane-inclined pad by supporting it with a pivot at the *centre of pressure* corresponding to the desired film thickness ratio. This aspect of the simplified theory has not been presented in sub-section 5.4.1 but it has been described elsewhere (Cameron, 1981).

A more general case is that in which a *known taper* is machined in a pad – i.e. only $h_2 - h_1$ is known at the outset. The technique required to deal with this practical situation is outlined in the following extended example.

### 5.5.4 Example of Calculations for a Thrust Bearing of Known Taper

An inclined thrust bearing of rectangular form, 90 mm wide by 30 mm long, is machined with a uniform taper of 0.03 mm over its length of 30 mm. It carries a load of 13 kN at a sliding speed of 10 m/s when using Tellus 68 oil supplied at an inlet temperature of 40°C. Determine the operating conditions of the bearing.

In this example neither the film thicknesses, the effective viscosity nor the temperature rise of the oil is known. For convenience, the enlarged viscosity–temperature characteristic of the relevant oil has been replotted in Figure 5.12 from Figure 5.2 over the temperature range 40–60°C.

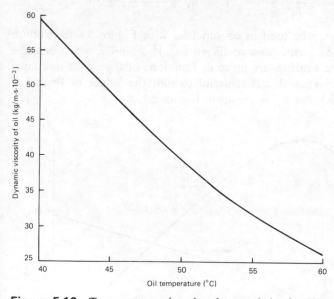

**Figure 5.12** Temperature–viscosity characteristics for Shell Tellus 68 oil. From data sheet on Tellus 68 oil:

| Temperature (°C) | $v$ (cSt) | $\eta$ (cP) | $\eta$ (kg/m·s) |
|---|---|---|---|
| 40 | 68 | 59.7 | 0.0597 |
| 45 | 56 | 49.2 | 0.0492 |
| 50 | 45 | 39.5 | 0.0395 |
| 55 | 36 | 31.6 | 0.0316 |
| 60 | 30 | 26.1 | 0.0261 |

**Table 5.2**

| $K$ | $h_1$ (mm) | $\bar{W}$ | $\bar{F}$ | $\bar{Q}$ | $\bar{T} = \bar{F}/\bar{Q}$ |
|------|------------|-----------|-----------|-----------|------------------------------|
| 0.5 | 0.06 | 0.1312 | 0.8437 | 0.6000 | 1.4062 |
| 0.75 | 0.04 | 0.1511 | 0.8028 | 0.6364 | 1.2615 |
| 1.00 | 0.03 | 0.1589 | 0.7726 | 0.6667 | 1.1589 |
| 1.25 | 0.024 | 0.1601 | 0.7488 | 0.6923 | 1.0816 |
| 1.50 | 0.02 | 0.1577 | 0.7292 | 0.7143 | 1.0209 |
| 2.00 | 0.015 | 0.1479 | 0.6972 | 0.7500 | 0.9296 |
| 2.50 | 0.012 | 0.1360 | 0.6711 | 0.7778 | 0.8626 |
| 3.00 | 0.010 | 0.1242 | 0.6484 | 0.8000 | 0.8105 |

Now the geometry of the oil film is described, by the variable $K$, where $K = (h_2 - h_1)/h_1$, but in this example only $h_2 - h_1$ is known as 0.03 mm, so we may write

$$K = \frac{h_2 - h_1}{h_1} = \frac{0.03 \times 10^{-3}}{h_1} \quad \text{or} \quad h_1 = \frac{0.03 \times 10^{-3}}{K} \text{ m}$$

This relationship may be used in conjunction with Figure 5.8 to produce Table 5.2. Thus, the *dimensionless* coefficients $\bar{Q}$, $\bar{W}$, $\bar{F}$ and $\bar{T}$, which describe the operation of the bearing, are given as functions of the as yet unknown *dimensional* film thickness $h_1$. It is useful to plot the values of $\bar{W}$ and $\bar{T}$ against the film thickness $h_1$ as shown in Figure 5.13.

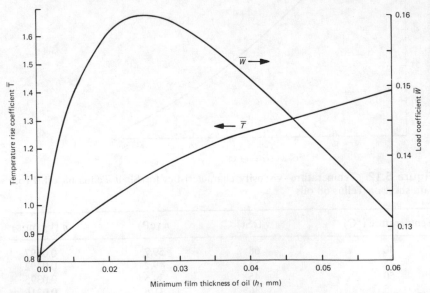

**Figure 5.13** Load and temperature coefficients as functions of minimum film thickness

Now the known values may be inserted into the third expression of (5.12), to give

$$13\,000 = \bar{W}\left\{\eta \times 10 \times 0.09\left(\frac{30 \times 10^{-3}}{h_1}\right)^2\right\}$$

or

$$\eta = 16.05 \times 10^6\left(\frac{h_1^2}{W}\right) \qquad (a)$$

Similarly, putting $k = 0.8$, $\rho = 878\ \text{kg/m}^3$ and $C_p = 1880\ \text{J/kg·°C}$ into (5.12) gives

$$\Delta T = 0.8 \times \bar{T}\left(\frac{\eta \times 10 \times 30 \times 10^{-3}}{878 \times 1880 \times h_1^2}\right)$$

$$\Delta T = 0.145 \times 10^{-6}\left(\frac{\eta \bar{T}}{h_1^2}\right) \qquad (b)$$

Again

$$T_e = T_i + \frac{\Delta T}{2} \qquad (c)$$

Thus, there are *six* unknown quantities to be found from the three derived expressions (a), (b) and (c) above, together with the three relationships given in Figures 5.12 and 5.13. The following procedure may be used to obtain solutions for these six relationships:

(1) Guess an initial value of $h_1$.
(2) Obtain values of $\bar{W}$ and $\bar{T}$ from Figure 5.13.
(3) Evaluate $\eta$ from expression (a).
(4) Evaluate $T_e$ from Figure 5.12.
(5) Evaluate $\Delta T$ from expression (b).
(6) Evaluate $T_e$ from expression (c).

Now, if the initially guessed value of $h_1$ is close to the actual value, then the effective temperature, $T_e$, found in step (4) above will be very close to the value of $T_e$ found from step (6). Usually further adjustment to $h_1$ will be necessary in order to obtain a consistent set of values for all of the six unknowns listed above.

Take an initial guess of $h_1 = 0.03\ \text{mm} = 0.03 \times 10^{-3}\ \text{m}$ and proceed through the six steps described above:

(1) $h_1 = 0.03 \times 10^{-3}\ \text{m}$
(2) $\bar{W} = 0.1589$ and $\bar{T} = 1.1589$ from Figure 5.15
(3) $\eta = 0.0909$ from expression (a)
(4) $T_e = 34°\text{C}$ from Figures 5.13 and 5.2

(5) $\Delta T = 16.97°C$ from expression (b)

(6) $T_e = 48.5°C$ from expression (c)

The values for the effective temperature, $T_e$, obtained in steps (4) and (6) are clearly inconsistent, and a trial will show that a reduction in the value of $h_1$ is necessary to cause convergence of these values. Subsequent results obtained by repeating the above set of calculations for various values of $h_1$ are shown in Table 5.3. The convergence is clearly sensitive to small changes in $h_1$ but the difference of only 0.56°C between the values of $T_e$ obtained in the fourth line of the table is sufficiently accurate. The bearing will therefore operate with a minimum film thickness, $h_1$, of 0.021 mm and a temperature rise, $\Delta T$, of 15.35°C, so that the effective temperature of the oil, $T_e$, becomes $\frac{1}{2}(47.1 + 47.66) = 47.4°C$. Reference to Figure 5.12 shows that the effective viscosity of the oil is 0.0443 kg/m·s.

The load capacity of the bearing may now be checked as follows from the third expression of (5.12):

$$W = 0.1586\left\{0.0443 \times 10 \times 0.09\left(\frac{30 \times 10^{-3}}{0.021 \times 10^{-3}}\right)^2\right\}$$

$$= 12\,905 \text{ N}$$

which is close to the given load of 13 000 N.

Again the temperature rise may be confirmed from expressions (5.12) as

$$\Delta T = 0.8 \cdot 1.04\left\{\frac{0.0443 \times 10 \times 30 \times 10^{-3}}{878 \times 1880(0.021 \times 10^{-3})^2}\right\}$$

$$= 15.2°C$$

and

$$T_e = 40 + \frac{15.2}{2} = 47.6°C$$

which is very close to the value of 47.4°C used above.

**Table 5.3**

| $h_1$ (mm) | $\bar{W}$ | $\bar{T}$ | $\eta$ (kg/m·s) | $T_e$ (°C) | $\Delta T$ (°C) | $T_e$ (°C) |
|---|---|---|---|---|---|---|
| 0.03 | 0.1589 | 1.1589 | 0.0909 | $\approx 34$ | 16.97 | 48.5 |
| 0.025 | 0.1600 | 1.1000 | 0.063 | $\approx 37$ | 16.0 | 48.0 |
| 0.023 | 0.1598 | 1.0700 | 0.053 | 43.1 | 15.58 | 47.8 |
| 0.021 | 0.1586 | 1.0400 | 0.045 | 47.1 | 15.35 | 47.66 |
| 0.020 | 0.1576 | 1.022 | 0.0408 | 49.4 | 15.10 | 47.55 |
| 0.0205 | 0.1581 | 1.030 | 0.0426 | 48.3 | 15.27 | 47.63 |

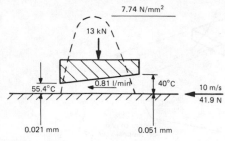

**Figure 5.14**

Remembering that $K = 0.030/0.021 = 1.428$ allows $\bar{Q}$ to be interpolated from Table 5.2 as 0.71 and the flow rate follows from the second expression of (5.12) as

$$Q = 0.71\{0.09(0.021 \times 10^{-3}) \times 10\} = 13.4 \times 10^{-6}\,\text{m}^3/\text{s}$$

$$= 0.81\ \text{l/min}$$

Similar calculations to those presented previously allow the maximum pressure in the bearing to be found as $7.74\ \text{N/mm}^2$ and the shear force as 42 N. The operating characteristics of the bearing may therefore be summarised by the diagram shown in Figure 5.14.

## 5.6   THRUST BEARINGS OF FINITE WIDTH

### 5.6.1   Two-dimensional Form of Reynolds's Equation

It was shown in sub-section 5.4.1 that the pressure distribution in a plane-inclined slider bearing of high width-to-length ratio, $B/L$, is given by the solution of the one-dimensional form of Reynolds's equation (5.7):

$$\frac{\partial}{\partial x}\left\{\frac{h^3}{\eta}\frac{\partial p}{\partial x}\right\} = 6U_0\frac{\partial p}{\partial x}$$

Inherent in this equation is the assumption that there is no flow of fluid in the direction of the $y$ axis, so that $\partial p/\partial y = 0$. Considering the inclined slider bearing of high $B/L$ ratio shown in Figure 5.15(a), it is clear that this assumption will be approximately true away from the ends of the bearing. However, near the ends of the bearing the lubricant is able to flow in the direction of the $y$ axis and to escape from the bearing in this direction; consequently, in these regions $\partial p/\partial y \neq 0$.

The effects of the flow of fluid in the direction of the $y$ axis, at right angles to the direction of sliding, is to diminish the pressures in the fluid at the ends of the bearing, as shown in Figure 5.15(b). Away from the ends of

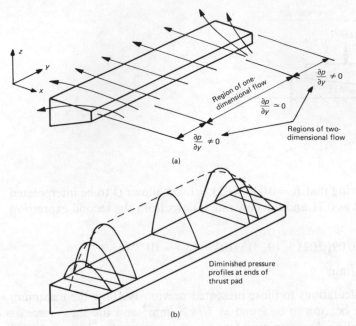

**Figure 5.15**

the bearing it has been assumed that the one-dimensional form of Reynolds's equation is applicable and that the load capacity, for example, may be calculated by neglecting the reductions in pressure which occur near the ends of the bearing. Effectively, this means that the pressure profile deduced from the one-dimensional form of Reynolds's equation is assumed to exist over the full width, $B$, of the bearing.

This is a remarkably good assumption, as in the case of a bearing of $B/L$ ratio equal to 2 the actual load capacity is typically only about 30% less than that calculated on the basis of one-dimensional theory. However, if a square bearing, of $B/L$ ratio equal to unity, is considered, then the flow of fluid in the direction of the $y$ axis may not be ignored and a two-dimensional solution is necessary. Such two-dimensional solutions are obtained from the following two-dimensional form of Reynolds's equation (Cameron, 1981):

$$\frac{\partial}{\partial x}\left\{\frac{h^3}{\eta}\frac{\partial p}{\partial x}\right\} + \frac{\partial}{\partial y}\left\{\frac{h^3}{\eta}\frac{\partial p}{\partial y}\right\} = 6U_0\frac{\partial h}{\partial x} \qquad (5.14)$$

The addition of the second term on the left-hand side involves derivatives with respect to $y$ and therefore includes the effects of flow in the direction of the $y$ axis.

The flow pattern of fluid within a square bearing will be of the form shown in Figure 5.16(a), while the corresponding pressure distribution will be as indicated in Figure 5.16(b). It will be noticed that there is no extended

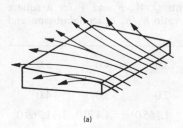 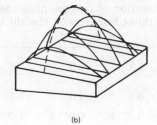

(a)                         (b)

**Figure 5.16**

central region in which $\partial p/\partial y = 0$, and as a consequence the maximum pressure generated in a square bearing is considerably lower than that in a wide bearing which operates under the same conditions.

### 5.6.2 Characteristics of a Square Bearing

Solution of the two-dimensional form of Reynolds's equation enables the important operating parameters of a square inclined slider bearing to be evaluated and expressed in the same form as that developed from first principles for a wide bearing in sub-section 5.4.2. Thus, the flow rate, $Q$, the load capacity, $W$, the shear force, $F$, and the temperature rise, $\Delta T$, for a square bearing will also be given by expressions (5.12) and (5.13), *provided that* the appropriate values of the non-dimensional coefficients $\bar{Q}$, $\bar{W}$, $\bar{F}$ and $\bar{T}$ are used. These expressions are restated below:

$$Q = \bar{Q}(Bh_1 U_0)$$

$$W = \bar{W}\left\{\eta U_0 B\left(\frac{L}{h_1}\right)^2\right\}$$

$$F = \bar{F}\left(\frac{\eta U_0 BL}{h_1}\right)$$

$$\Delta T = k\bar{T}\left(\frac{\eta U_0 L}{\rho C_p h_1^2}\right)$$

Again the dimensionless coefficients depend only on the geometry of the film as described by the film thickness ratio $h_2/h_1$ or the parameter $K$. Values of the four coefficients for a square bearing are tabulated in Table 5.4 and plotted in Figure 5.17 as functions of $h_2/h_1$ and $K$. These results were obtained by numerical analysis (Jakobsson and Floberg, 1958) and should be compared with the results for a wide bearing given in Table 5.1 and Figure 5.8.

It now follows that a square bearing may be analysed and designed with the aid of the above four expressions and Figure 5.8, *without any detailed knowledge* of how the values of the four coefficients were obtained.

**Table 5.4** Variation of dimensionless coefficients $\bar{Q}$, $\bar{W}$, $\bar{F}$ and $\bar{T}$ for a square hydrodynamic thrust bearing with film thickness ratio $h_2/h_1$ (after Jakobsson and Floberg, 1958)

| $\dfrac{h_2}{h_1}$ | 1 | 1.5 | 2.0 | 2.5 | 3.0 | 4.0 | 5.0 |
|---|---|---|---|---|---|---|---|
| $K$ | 0 | 0.5 | 1.0 | 1.5 | 2.0 | 3.0 | 4.0 |
| $\bar{Q}$ | 0.5000 | 0.6796 | 0.8473 | 1.0080 | 1.1650 | 1.4700 | 1.7690 |
| $\bar{W}$ | 0 | 0.05575 | 0.06894 | 0.06997 | 0.06701 | 0.05836 | 0.05011 |
| $\bar{F}$ | 1.0000 | 0.8249 | 0.7276 | 0.6633 | 0.6163 | 0.5496 | 0.5026 |
| $\bar{T}$ | 2.0000 | 1.2140 | 0.8588 | 0.6579 | 0.5291 | 0.3739 | 0.2842 |

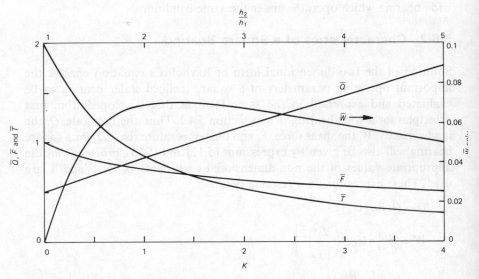

**Figure 5.17** Variation of dimensionless coefficients $\bar{Q}$, $\bar{W}$, $\bar{F}$ and $\bar{T}$ for a square hydrodynamic thrust bearing with film thickness ratio $h_2/h_1$

Similar sets of data have been obtained for bearings of other width-to-length ratios, but as most of the individual bearings used in practical designs of thrust bearings are of an approximately square form, the information presented in Table 5.4 and Figure 5.17 is particularly useful.

### 5.6.3 Example of Calculations for a Square Bearing

A thrust bearing of square form works with a uniform sliding velocity of 56 m/s and supports a steady load of 25 kN. The bearing is to be lubricated with Tellus 68 oil, which is available at an inlet temperature of 40°C. The bearing must be designed to meet the following three conditions: (1) the minimum film thickness is not to be less than 0.025 mm; (2) the flow rate of oil is not to exceed 10 l/min; and (3) the outlet temperature of the oil is not to exceed 100°C.

In order to assign an initial value to the length and width of the bearing, it is remembered that the mean pressure in the bearing of sub-section 5.5.4 was about 4.8 N/mm$^2$, so that in the present case

$$L = B = \left(\frac{25\,000}{4.8}\right)^{\frac{1}{2}} = 72 \text{ mm}$$

Thus, as a conservative guess, the initial value of $L$ and $B$ is taken to be 80 mm. Study of Figure 5.17 shows that the value of $\bar{W}$ in the region of $K = 1.5$ does change greatly from the tabulated value of 0.069 97, with the corresponding value of $\bar{T} = 0.6579$. These two values will be taken as constant for the purposes of this example.

Now substitute all known values in the expressions for $W$ and $\Delta T$ from expressions (5.12) and (5.13):

$$25 \times 10^3 = 0.069\,97 \left\{ \eta 56 \times 80 \times 10^{-3} \left(\frac{80 \times 10^{-3}}{h_i}\right)^2 \right\}$$

or

$$12.461 \times 10^6 = \frac{\eta}{h_1^2} \tag{a}$$

Again

$$\Delta T = 0.8 \times 0.6579 \left\{ \frac{\eta 56 \times 80 \times 10^{-3}}{880 \times 1880 h_1^2} \right\}$$

$$\Delta T = 1.425 \times 10^{-6} \left(\frac{\eta}{h_1^2}\right) \tag{b}$$

Substitute for $\eta/h_1^2$ from (b) into (a), to give

$$\Delta T = 1.425 \times 10^{-6} \times 12.461 \times 10^6 = 17.8°C$$

The effective temperature, $T_e$, follows as

$$T_e = 40 + \frac{17.8}{2} = 48.9°C$$

and the outlet temperature, $T_o$, follows as 57.8°C.

The effective viscosity of Tellus 68 at the temperature of 48.9°C may be obtained from Figure 5.13 as 0.0413 kg/m·s. Substitution of this value into

(a) allows the value of the minimum film thickness, $h_1$, to be found:

$$h_1^2 = \frac{0.0413}{12.461 \times 10^6} \quad \text{or} \quad h_1 = 5.75 \times 10^{-5}\,\text{m} \quad \text{or} \quad 0.058\,\text{mm}$$

From Figure 5.12, $\bar{Q} = 1.008$ at $K = 1.5$ and the flow rate, $Q$, follows from expressions (5.12) as

$$Q = 1.008(80 \times 10^{-3} \times 0.058 \times 10^{-3} \times 56)$$

$$= 0.2619 \times 10^{-3}\,\text{m}^3/\text{s} = 15.71\,\text{l/min}$$

This flow rate is in excess of the allowed value of 10 l/min and the choice of $L = B = 80$ mm is clearly unsatisfactory. By taking three additional values of $L = 40$ mm, 50 mm and 65 mm and repeating the calculations outlined above, Table 5.5 may be constructed. As limits have been set on $h_1$, $Q$ and $T_o$ in the design specification, the results of Table 5.5 were plotted in Figure 5.18 to show the variation of these critical quantities with the length and width of the bearing.

Now the limitations which have been set on $h_1$, $Q$ and $T_o$ may be superimposed on the curves of Figure 5.18, to give the bounded characteristics shown.

(1) As $h_1$ is limited to a minimum value of 0.025 mm, the horizontal line A is drawn through the curve for $h_1$ at a value of 0.025 mm.
(2) As $Q$ is limited to a maximum value of 10 l/min, the line B is drawn through the curve for $Q$.
(3) As $T_o$ is limited to a maximum value of 100°C, the line C is drawn through the curve for $T_o$.

Study of Figure 5.18 reveals that:

(1) Bearings of length less than 53.5 mm operate with values of $h_1$ less than 0.025 mm.
(2) Bearings of length greater than 69.3 mm require flow rates in excess of 10 l/min.
(3) Bearings of length less than 44 mm generate outlet temperatures greater than 100°C.

**Table 5.5**

| $L$ (mm) | $\Delta T$ (°C) | $\eta$ (kg/m·s) | $h_1$ (mm) | $Q$ (l/min) | $T_o$ (°C) |
|---|---|---|---|---|---|
| 80 | 17.8 | 0.0413 | 0.058 | 15.7 | 57.8 |
| 65 | 26.9 | 0.0326 | 0.037 | 8.23 | 66.9 |
| 50 | 45.6 | 0.0239 | 0.022 | 3.72 | 85.6 |
| 40 | 71.0 | 0.0156 | 0.012 | 1.69 | 111.0 |

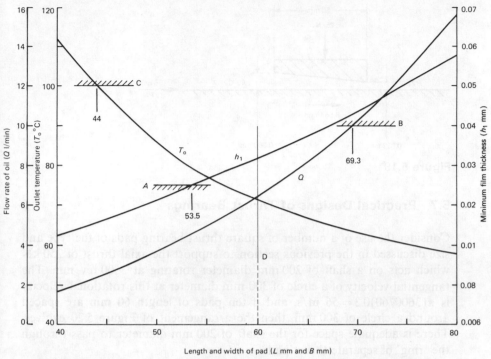

**Figure 5.18**  Bounded characteristics of a square bearing using Tellus 68 oil

It is, therefore, clear from Figure 5.18 that an acceptable design of bearing must have a length of between 53.5 mm, line A, and 69.3 mm, line B. If the boundary of line A is not violated, then that of line C is also satisfied.

As the relative importance of the limits set on $h_1$ and $Q$, by lines A and B, has not been stated, it is reasonable to select a length of bearing approximately midway between the lengths corresponding to lines A and B – say 60 mm, as indicated by line D in the figure.

The values of $h_1$, $Q$ and $T_o$ corresponding to a length of 60 mm may be read directly off Figure 5.18 or recalculated as a check in the manner employed previously. It then follows that $h_1$ has a value of 0.032 mm, $Q$ a value of 6.56 l/min and $T_o$ a value of 71.6°C. The table of Figure 5.22 gives the coefficient $\bar{F}$ as 0.6633, so that the shear force becomes, from expressions (5.12),

$$F = 0.6633\left(\frac{0.0308 \times 56 \times 0.06 \times 0.06}{0.032 \times 10^{-3}}\right)$$

$$= 127 \text{ N}$$

and the power consumed by the bearing is simply $FU_o = 127 \times 56 = 7112$ W.
The operating conditions of the bearing are shown in Figure 5.19.

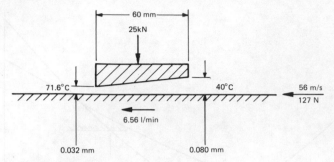

**Figure 5.19**

## 5.7 Practical Designs of Thrust Bearings

Consider the use of a number of square thrust bearing pads of the type and size discussed in the previous section to support the axial thrust of 250 kN which acts on a shaft of 200 mm diameter rotating at 3600 rev/min. The tangential velocity of a circle of 300 mm diameter at this rotational velocity is $\pi(3600/60)0.3 = 56$ m/s, and if ten pads of length 60 mm are spaced around a circle of 300 mm, then the arrangement of Figure 5.20 evolves. There is adequate space for the shaft of 200 mm diameter to pass through the 'ring' of separate thrust pads.

As ten pads of the final design deduced in sub-section 5.6.3 are used in the complete bearing of Figure 5.20, its specification is as follows:

Load capacity $= 25 \times 10^3 \times 10 = 250$ kN
Flow rate of oil $= 6.56 \times 10 = 65.6$ l/min
Inlet temperature $= 40°C$
Outlet temperature $= 71.6°C$
Lubricant $=$ Tellus 68 oil
Shaft velocity $= 3600$ rev/min
Mean velocity over pads $= 56$ m/s
Power consumed $= 7112 \times 10 = 71$ kW
Minimum film thickness $= 0.032$ mm
Length and width of pads $= 60$ mm

The analysis presented here is typical of the methods used in bearing design, but the schematic arrangement shown in Figure 5.20 is a gross simplification, and details of the construction of more realistic bearings are shown in Figure 5.21. An important feature of practical thrust bearings is that the necessary convergent geometry of the pads is created by allowing them to tilt about a fixed pivot rather than by attempting to build a known taper into the assembly. Consequently, the analysis and design of such practical thrust bearings is more complicated than outlined here, and more details are available elsewhere (Martin, 1970).

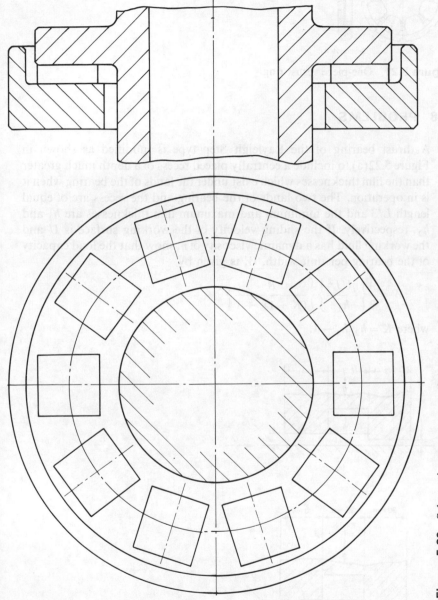

**Figure 5.20** Schematic arrangement of thrust bearing

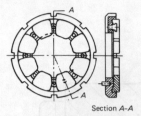

Section A-A

**Figure 5.21** One-piece thrust ring

## 5.8 PROBLEMS

**1** A thrust bearing of the Rayleigh Step type is modified as shown in Figure 5.22(a) to include a centrally placed recess of a depth much greater than the film thicknesses which exist under the lands of the bearing when it is in operation. The two lands of the bearing and the recess are of equal length $L/3$ and the minimum and maximum film thicknesses are $h_1$ and $h_2$, respectively. If the sliding velocity of the working surface is $U$ and the working fluid has a dynamic viscosity of $\eta$, show that the load capacity of the bearing per unit width, $W$, is given by

$$W = \frac{4}{3} \left\{ \frac{\eta U L^2}{h_1^2} \right\} \left\{ \frac{K}{(K+1)^3 + 1} \right\}$$

where $K = h_2/h_1 - 1$.

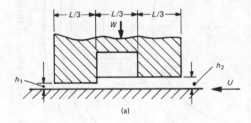

(a)

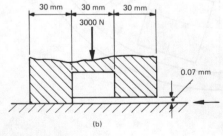

(b)

**Figure 5.22**

A bearing of this type lies with one of its lands in contact with its stationary working surface, as shown in Figure 5.22(b), under a load of 3000 N. The gap between the other land and the stationary surface is 0.07 mm, the length of the bearing is 90 mm and its width is 240 mm. If the working fluid has a dynamic viscosity of 0.05 kg/m·s, estimate the sliding velocity necessary to separate the bearing from the working surface. [0.15 m/s]

2   The inclined hydrodynamic bearing pad shown in Figure 5.23(a), of length $L$ and width $B$, operates with a fluid of dynamic viscosity $\eta$ and is positioned relative to a plane slider, which has a velocity $U$ in the direction shown, so that the film thickness, $h_2$, at entry is $k$ times the film thickness, $h_1$, at exit. During operation the pressure in the fluid at exit is maintained at a value of $p_1$ avoe atmospheric, while that at entry, $p_2$, remains at the atmospheric value. On the assumption that the width, $B$, is much greater than the length, $L$, solution of Reynolds's equation shows that the load capacity of the bearing, $W$, is given by the expression

$$W = \alpha_1 \left\{ \eta U B \left( \frac{L}{h_1} \right)^2 \right\} + \alpha_2 \{ BLp_1 \}$$

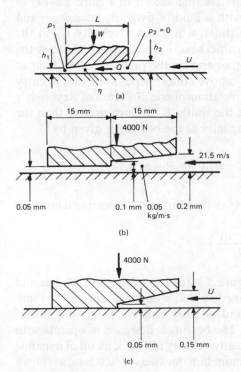

Figure 5.23

and the volumetric flow rate, $Q$, of fluid through the bearing is given by the expression

$$Q = \alpha_3 \{ BUh_1 \} - \alpha_4 \left\{ \frac{Bh_1^3 p_1}{\eta L} \right\}$$

When $k = 2$, $\alpha_1 = 0.1589$, $\alpha_2 = \frac{1}{3}$, $\alpha_3 = \frac{2}{3}$ and $\alpha_4 = \frac{2}{9}$; when $k = 3$, $\alpha_1 = 0.1479$, $\alpha_2 = \frac{1}{4}$, $\alpha_3 = \frac{3}{4}$ and $\alpha_4 = \frac{3}{8}$; where $k = h_2/h_1$.

(a) The thrust pad shown in Figure 5.23(b) has a length of 30 mm and a width of 100 mm, and is machined so that a uniform taper of 0.1 mm exists over a length of 15 mm, while a parallel step of height 0.05 mm exists over the remaining length of 15 mm. A load of 4000 N is carried by the bearing when the slider has a velocity of 21.5 m/s and an oil of dynamic viscosity 0.05 kg/m·s is used. Show that the bearing will operate with a minimum film thickness of 0.05 mm, as shown in Figure 5.23(b).

(b) If the load of 4000 N is imposed on the bearing when it starts from rest, as shown in Figure 5.23(c), estimate the sliding velocity at which the bearing will separate from the sliding member.
[13.72 m/s]

3   The inclined hydrodynamic thrust bearing shown in Figure 5.24(a) of length $L$ and width $B$ operates with a fluid of dynamic viscosity $\eta$ and is positioned relative to a plane slider, which has a velocity $U_0$ in the direction shown, so that the film thickness, $h_2$, at entry is 1.5 times the film thickness, $h_1$, at exit. During operation the pressure in the fluid at exit is maintained at a value of $p_1$ above atmospheric, while that at entry is maintained at a value of $p_2$ above atmospheric. Solution of Reynolds's equation on the assumption that the width, $B$, is much greater than the length, $L$, shows that the load capacity of the bearing is given by

$$W = 0.1312 \left\{ \eta U_0 B \left( \frac{L}{h_1} \right)^2 \right\} + BL \left( \frac{2}{5} p_1 + \frac{3}{5} p_2 \right)$$

and that the volumetric flow rate, $Q$, of fluid through the bearing is given by

$$Q = \frac{3}{5}(U_0 B h_1) + \frac{3Bh_1^3}{20\eta} \left( \frac{p_2 - p_1}{L} \right)$$

The thrust bearing shown in Figure 5.24(b) has a length of 50 mm and a width of 200 mm, and is machined so that a uniform taper of 0.04 mm exists over a length of 25 mm and another uniform taper of 0.06 mm exists over the remaining 25 mm. The bearing is designed to operate with a plane slider which moves at a steady velocity of 4 m/s, an oil of dynamic viscosity 0.08 kg/m·s and a minimum film thickness of 0.08 mm, as shown in Figure 5.23(b).

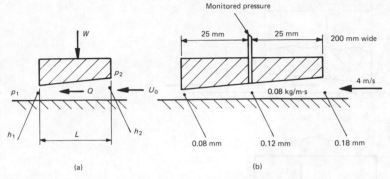

**Figure 5.24**

It is proposed to monitor the operation of the bearing by measuring the pressure at the mid-section of the bearing, where the film thickness is expected to be 0.12 mm. Find the value of this pressure for the configuration shown in Figure 5.24(b), and, hence, estimate the total load capacity of the bearing.
$[0.571 \times 10^6 \text{ N/m}^2, 4040 \text{ N}]$

4 The inclined hydrodynamic bearing pad shown in Figure 5.25(a), of length $L$ and width $B$, operates with a fluid of dynamic viscosity $\eta$ and is positioned relative to a plane slider, which has a velocity $U$ in the direction shown, so that the film thickness, $h_2$, at entry is twice the film thickness, $h_1$, at exit. During operation the pressure in the fluid at exit is maintained at a value of $p_1$ above atmospheric, while that at entry, $p_2$, remains at the atmospheric value. Solution of Reynolds's equation, on the assumption that the width, $B$, is much greater than the length, $L$, shows that the load capacity, $W$, of the bearing is given by

$$W = 0.1589 \left\{ \eta U B \left( \frac{L}{h_1} \right)^2 \right\} + \frac{1}{3}(BLp_1)$$

and that the volumetric flow rate, $Q$, of fluid through the bearing is given by

$$Q = \frac{2}{3}(BUh_1) - \frac{2}{9}\left( \frac{Bh_1^3 p_1}{\eta L} \right)$$

The mineral dressing mill shown schematically in Figure 5.25(b) weighs $8 \cdot 10^6$ N, rotates about the vertical axis at 360 rev/min and is supported on 49 identical thrust pads which are equispaced around the circumference of an annular surface of the mill at a mean diameter of 2 m. Each pad has a length of 90 mm and a width of 240 mm and is machined so that a uniform taper of 0.07 mm exists over a distance of 60 mm, as shown in Figure 5.25(c).

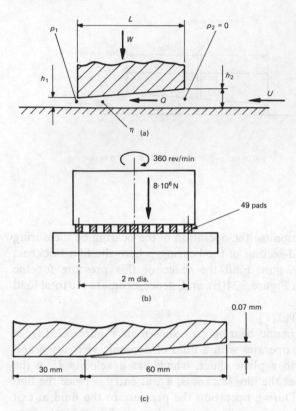

**Figure 5.25**

Three different oils of dynamic viscosity 0.02, 0.04 and 0.06 kg/m·s are available. Select one of these oils so that the pads operate with a minimum film thickness of about 0.07 mm.

[0.06 kg/m·s]

# Chapter 6
# Fluid-lubricated Journal Bearings

## 6.1 INTRODUCTION

In Chapter 5 bearings of plane form only were considered and these were shown to be of use in supporting loads which act at right angles to their plane. Bearings of this type are generally recognised as thrust bearings. However, another important type of bearing is the journal bearing. Journal bearings are used to support rotating circular shafts which are loaded in a radial direction.

In many applications it is found that a hydrodynamic journal bearing is most suitable for the range of loads and speeds involved, and it is fortunate that the shape of the oil film which is formed between the bore of a circular bearing and an eccentric rotating shaft is conducive to the generation of hydrodynamic pressures which are, in practice, able to support considerable radial loads.

### 6.1.1 Geometry

Consider the geometry of the mutually eccentric circles of similar diameter shown in Figure 6.1. In this figure the outer circle of radius $OA$, equal to $R_1$, represents the bore of the journal bearing and the inner circle of slightly smaller radius $CB$, equal to $R_2$, represents the periphery of the rotating circular shaft. The shaft is assumed to rotate clockwise with a peripheral velocity $U_0$ within the journal bearing, as shown in Figure 6.1. The eccentricity between the shaft and the journal bearing is represented by the distance, $e$, between the centre of the journal $O$ and the centre of the shaft $C$.

In the study of journal bearings the straight line $EOCF$ is known as the 'line of centres' and the distance $CO$ as the eccentricity of the shaft in the journal bearing.

161

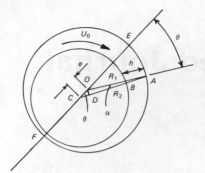

**Figure 6.1** Basic geometry of a journal bearing

It is required to obtain an expression for the varying distance, $h$, which exists between the surfaces of the shaft and journal bearing at a location given by the angle $\theta$, which is measured clockwise to $A$ from the line of centres at location $E$, where the maximum value of $h$ exists. In an operating hydrodynamic journal bearing, the space between the two surfaces is filled with a fluid and $h$ will therefore correspond to the varying film thickness in the bearing.

The perpendicular $OD$ is constructed from $O$ onto $AC$ and it is clear that

$$AC = R_2 + h = e \cos \theta + R_1 \cos \alpha$$

where $\alpha$ is the angle $OAC$. From the triangle $OCA$ it follows that

$$OD = (CO) \sin \theta = (AO) \sin \alpha$$

or

$$\sin \alpha = \frac{(CO)}{(AO)} \sin \theta = \left(\frac{e}{R_1}\right) \sin \theta$$

In practical journal bearings $e/R_1$ is about $10^{-3}$, so that the above expression for $\cos \alpha$ may be expressed as

$$\cos \alpha = \left\{1 - \left(\frac{e}{R_1}\right)^2 \sin^2 \theta\right\}^{\frac{1}{2}} = 1 - \frac{1}{2}\left(\frac{e}{R_1}\right)^2 \sin^2 \theta$$

Put this identity in the above expression for $AC$, to obtain

$$AC = R_2 + h = e \cos \theta + R_1 \left\{1 - \frac{1}{2}\left(\frac{e}{R_1}\right)^2 \sin^2 \theta\right\}$$

or

$$R_2 + h = e \cos \theta + R_1$$

as the term involving $\sin^2 \theta$ is very much less than unity. So

$$h = (R_1 - R_2) + e \cos \theta$$

Now $R_1 - R_2$ is the difference between the radii of the journal bearing and of the shaft. This difference is also recognised as the uniform radial clearance between the bearing and the shaft in the concentric case. Putting $R_1 - R_2 = c$ gives the following expression for $h$:

$$h = c + e \cos \theta$$

When $\theta = 0$, $h = c + e$, the maximum film thickness at $E$, and when $\theta = \pi$, $h = c - e$, the minimum film thickness at $F$. Conventionally, $h$ is expressed as follows:

$$h = c \left\{ 1 + \left( \frac{e}{c} \right) \cos \theta \right\}$$

or

$$h = c(1 + \varepsilon \cos \theta) \tag{6.1}$$

where $\varepsilon = e/c$ is known as the 'eccentricity ratio'. This is a dimensionless geometric parameter which describes completely the shape of the fluid film which exists between the surfaces of the journal bearing and shaft.

### 6.1.2   Hydrodynamic Action in Journal Bearings

Remembering that the radial clearance, $c$, and therefore the film thickness, $h$, are very much smaller than the radius of either the journal bearing or the shaft, then the shape of the film between the two surfaces may be developed into the equivalent orthogonal form shown in Figure 6.2(a). The clockwise rotation of the shaft shown in Figure 6.1 produces a tangential surface velocity, $U_0$, which appears as the velocity of the lower surface in Figure 6.2(a). In Figure 6.2(a) the upper surface is fixed in space and corresponds to that of the journal bearing. The coordinate $\theta$ varies from zero, where the film thickness is $c + e$, to $\pi$, where the film thickness is $c - e$, and then to $2\pi$, where the film thickness is again $c + e$.

Inspection of Figure 6.2(a) shows that a convergent film exists between $\theta = 0$ and $\theta = \pi$, while a divergent film exists between $\pi$ and $2\pi$. As the direction of sliding of the lower surface, at a velocity $U_0$, is in the direction of convergence between $\theta = 0$ and $\theta = \pi$, it is clear that hydrodynamic pressures will be developed in the fluid as it flows through the bearing. This important deduction follows from the general nature of the results which were obtained in Chapter 5 for the plane convergent geometry of Figure 5.7.

It must be remembered that the sections corresponding to $\theta = 0$ and $\theta = 2\pi$ in Figure 6.2(a) do, of course, define the same section because of the actual circular shape shown in Figure 6.1.

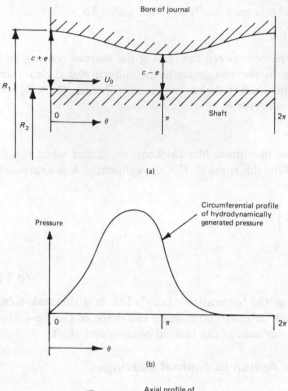

(a)

(b)

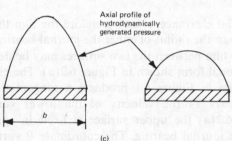

(c)

**Figure 6.2** Film shape and pressure profile between shaft and bore of a hydrodynamic journal bearing

The existence of the combined 'convergent–divergent' geometry of the film shown in Figure 6.2(a) produces a hydrodynamic pressure profile of the general shape given in Figure 6.2(b). It is of importance that the generation of pressure extends into the divergent position of the film beyond the section where the film thickness has its minimum value at $\theta = \pi$.

Now the profile shown in Figure 6.2(b) is typical of the variation of hydrodynamically generated pressure in the circumferential direction. However, as most bearings have an axial length $b$ which is usually less than their

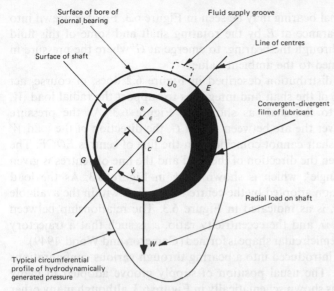

Surface of bore of journal bearing

Fluid supply groove

Surface of shaft

Line of centres

$U_0$

E

Convergent-divergent film of lubricant

Radial load on shaft

Typical circumferential profile of hydrodynamically generated pressure

**Figure 6.3** Operation of a hydrodynamic journal bearing

diameter, a considerable flow of fluid occurs in the axial direction, so that pressure profiles of the shape given in Figure 6.2(c) exist in the axial direction. The actual shape of the hydrodynamic pressure distribution in a journal bearing is represented by a combination of the pressure profiles given in Figure 6.2(b) and (c).

It may appear that the fluid in the bearing simply recirculates under the action of shaft rotation, as the sections at $\theta = 0$ and $2\pi$ represent the same actual section in Figure 6.1. However, complete recirculation of fluid does not occur, as some fluid is able to escape from the bearing by flowing in the axial direction. A bearing which completely encircles a shaft as shown in Figure 6.3 is known as a complete journal bearing.

It is, therefore, necessary to replace continually the fluid which escapes in the axial direction from the clearance by introducing replacement fluid into the bearing in the divergent region between $\theta = \pi$ and $\theta = 2\pi$.

The presence of the converging–diverging geometry and the need to replace fluid within the completely circular shape of the typical journal bearing introduce complications which were not present in the simple case of plane thrust bearings. These and other important aspects of journal bearings are considered further below.

### 6.1.3 Operation of Practical Journal Bearings

By combining the geometry of Figure 6.1 with the deduced pressure profile shown in Figure 6.2(b), the general features associated with the operation of a practical journal bearing may be seen in Figure 6.3. Fluid is drawn into the converging clearance at $E$ by the rotating shaft and some of this fluid flows completely through the bearing, to emerge at $G$, where the pressure in the fluid has returned to the ambient value.

The pressure distribution described in Figure 6.2 does, of course, act around the surface of the shaft and integrates to support the radial load, $W$, which is applied to the shaft as shown in Figure 6.3. As the pressure distribution acts over the arc between $E$ and $G$, the direction of the load $W$ which acts on the shaft cannot coincide with the line of centres $EOCF$. The relationship between the direction of the load and the line of centres is given by the 'attitude angle', which is shown as $\psi$ in Figure 6.3. As the load increases, the position adopted by the centre of the shaft within the available radial clearance, $c$, is as indicated in Figure 6.3. The relationship between the attitude angle, $\psi$, and the eccentricity ratio, $\varepsilon$, is such that a trajectory of approximately semicircular shape is formed (Cameron and Wood, 1949).

Fluid may be introduced into a bearing through various configurations of supply grooves. The usual position of supply groove adopted in many common bearings is shown schematically in Figure 6.3, although many other variations exist (Martin, 1983).

The analysis of the flow of fluid within the confines of a journal bearing is clearly much more complicated than in the case of plane thrust bearings. Consequently, a two-dimensional form of Reynolds's equation must be solved (Cameron, 1981), and additional complications are encountered in the establishment of the correct boundary conditions to be applied at the termination of the pressure profile in the divergent part of the clearance, section $G$ in Figure 6.3.

As, in general, flow of fluid into a convergent clearance causes an increase in pressure, it follows that its continued flow into a divergent clearance will lead to a decrease in pressure. However, when the pressure approaches the ambient value, air may be drawn into the divergent clearance, as there will be insufficient fluid available to fill the clearance, which increases in the direction of shaft rotation. Similarly, dissolved air or other gases may come out of solution in the fluid as the pressure decreases. Again, if the pressure falls below the vapour pressure of the fluid, then cavities composed of the fluid's vapour will form in the film. It is, of course, physically impossible for the pressure in the film to fall below zero absolute (Dowson and Taylor, 1975).

These physical limitations are generally referred to as cavitation, and result in streamers or striations of fluid being formed in a mixture of air and vapour in the divergent clearance (Cole and Hughes, 1956). Much effort has resulted in useful sets of boundary conditions being postulated (Dowson and Miranda, 1975; Etsion and Pinkus, 1975).

A more complete description of a typical complete journal bearing is given in Figure 6.4, where a shaft of diameter $d$ is supported in a bearing of axial length $b$. The shaft rotates with an angular velocity $N$ and supports a

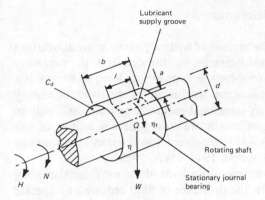

**Figure 6.4** Nomenclature of a hydrodynamic journal bearing

load $W$, while a power $H$ is required to rotate the shaft against the viscous resistance which is produced in the clearance of the bearing. Fluid is supplied at a flow rate $Q$ with viscosity $\eta_f$ through a supply groove of axial length $l$ and width $a$, which is machined in the wall of the bearing, but the bearing operates with an effective viscosity $\eta$. The difference between the diameters of the shaft and the journal bearing is $C_d$, so that the previously defined radial clearance, $c$, is simply $C_d/2$.

Other important parameters associated with the arrangement of Figure 6.4 are listed in the Notation section (pp. viii–xii) and introduced in the following sections.

## 6.2 SOLUTIONS FOR HYDRODYNAMIC JOURNAL BEARINGS

It is an experimentally observed fact that the behaviour of a hydrodynamic journal bearing may be described by using an effective or mean viscosity to take account of the variation of temperature and, hence, viscosity which occurs within the bearing under steady running conditions.

The effective temperature is determined by the flow rate of fluid, which is itself dependent on the pressure in the supply groove, the form and dimensions of the supply groove, the viscosity of the fluid, the speed of the shaft and the geometry of the bearing.

A certain minimum flow rate of fluid will be required for the bearing to form an adequate load-carrying film of full extent. In the design procedure outlined here, the flow rate required by the bearing is first determined, and the dimensions of the supply groove and pressure of fluid in it are then selected to provide the required flow rate.

### 6.2.1 Dimensionless Coefficients

In principle, the analysis of the behaviour of hydrodynamic journal bearings is the same as that of plane thrust bearings, but the details of the necessary mathematical analysis required to solve the appropriate form of Reynolds's equation leads to great complication. Inclusion of realistic boundary conditions makes it essential that numerical methods be used, and only in the case of bearings of very short length are closed-form solutions of any practical use (Dubois and Ocvirk, 1953; Woolacot, 1965): many contributions have been made (Cameron and Wood, 1949; Hays, 1959).

The important operating variables for a hydrodynamic journal bearing are the radial load supported, $W$, the flow rate of fluid required to operate successfully, $Q$, and the power required to rotate the shaft against the viscous resistance of the bearing, $H$. It is found that these variables may be expressed as follows:

$$W = \bar{W}\left\{\eta Nbd\left(\frac{d}{C_d}\right)^2\right\} \tag{6.2a}$$

$$Q = \bar{Q}(NbdC_d) \tag{6.2b}$$

$$H = \bar{H}\left(\eta\frac{N^2d^3b}{C_d}\right) \tag{6.2c}$$

Therefore, it is seen that, within acceptable accuracy, the load capacity, required flow rate and power consumption of a hydrodynamic journal bearing may be expressed as products of dimensionless coefficient and combinations of each bearing's dimensional features. The dimensionless coefficients, $\bar{W}$, $\bar{Q}$, $k$ and $\bar{H}$ have been shown to be functions of the geometry or shape of the fluid film under a particular set of conditions. Furthermore, for a complete journal bearing only two geometric variables are required to determine the values of the various dimensionless coefficients. These variables are the length-to-diameter ratio of the bearing, $b/d$, and its operating eccentricity ratio, $\varepsilon$.

These results should be compared with the similar solutions presented for thrust bearings of square form in sub-section 5.6.2.

Again, as in sub-section 5.4.5, a power balance for a journal bearing may be written down as follows:

$$\rho C_p Q\Delta T = kH$$

or

$$\Delta T = k\bar{T}\left\{\frac{\eta N}{\rho C_p}\left(\frac{d}{C_d}\right)^2\right\} \tag{6.2d}$$

where $\bar{T} = \bar{H}/\bar{Q}$ as in sub-section 5.4.5.

It now follows that a complete hydrodynamic journal bearing may be analysed and designed with the aid of the above four expressions without any detailed knowledge of how the values of the four dimensionless coefficients were derived. Values of these coefficients may be found in published work (Cameron and Wood, 1949; Jakobsson and Floberg, 1957), but some variations in the form of the dimensional groups have led to differing values of the coefficients given above.

## 6.2.2 Example of Journal-bearing Calculation

A hydrodynamic journal bearing of length 160 mm operates with a shaft of 200 mm diameter which rotates at 1200 rev/min. If the effective viscosity of the lubricating oil is assumed for simplicity to be 40 cP and the diametral clearance is 0.18 mm, determine the operating characteristics of the bearing at an eccentricity ratio of 0.7.

In this case the $b/d$ ratio of the bearing is seen to be $160/200 = 0.8$. References then show that the important dimensionless coefficients for a bearing of this length-to-diameter ratio at an eccentricity ratio of 0.7 are as follows: $\bar{W} = 9.9$, $\bar{Q} = 1.20$ and $\bar{H} = 79$. The value of viscosity to be used in the calculations may be found in SI units from expression (5.2) as $40/1000 = 0.04$ kg/m·s.

As in sub-section 5.4.5, the oil density will be taken as 880 kg/m³, the specific heat as 1880 Nm/kg·°C and the proportion of heat convected out of the bearing by the oil as 0.8. The consistent measure of angular velocity is revolutions per unit time, which in this case is $1200/60 = 20$ rev/s.

*Load capacity*

With $\bar{W} = 9.9$ in expression (6.2a),

$$W = 9.9\left\{0.4 \times 20(160 \times 10^{-3})(200 \times 10^{-3})\left(\frac{200 \times 10^{-3}}{0.18 \times 10^{-3}}\right)^2\right\}$$

$$W = 312\,900 \text{ N}$$

*Flow rate*

With $\bar{Q} = 1.2$ in expression (6.2b),

$$Q = 1.2\{20 \times (160 \times 10^{-3})(200 \times 10^{-3})(0.18 \times 10^{-3})\}$$

$$Q = 0.138 \times 10^{-3} \text{ m}^3/\text{s} = 8.29 \text{ l/min}$$

*Power*

With $\bar{H} = 79$ in expression (6.2c),

$$H = 79\left\{0.04 \times 20(200 \times 10^{-3})\frac{(160 \times 10^{-3})}{(0.18 \times 10^{-3})}\right\}$$

$$H = 8988 \text{ W} = 12 \text{ h.p.}$$

*Temperature rise*

With $k = 0.8$ and $\bar{T} = \bar{H}/Q = 79/1.2 = 65.833$ in expression (6.2d).

$$\Delta T = 0.8 \times 65.833\left\{\frac{0.04 \times 20}{880 \times 1880}\left(\frac{200 \times 10^{-3}}{0.18 \times 10^{-3}}\right)^2\right\}$$

$$\Delta T = 31.4°C$$

*Minimum film thickness*

With $\varepsilon = 0.7$ and $c = C_d/2 = (0.18 \times 10^{-3})/2 = 0.09 \times 10^{-3}$ in expression (6.1),

$$h_{min} = 0.09 \times 10^{-3}(1 + 0.7 \times \cos \pi)$$

$$h_{min} = 0.027 \times 10^{-3} \text{ m} = 0.027 \text{ mm}$$

## 6.3 ALTERNATIVE FORMS OF DIMENSIONLESS COEFFICIENTS

It is clear from the previous section that the operating characteristics of a hydrodynamic journal bearing may be determined easily, provided that the appropriate values of the dimensionless coefficients are known. However, the values of the dimensionless coefficients may only be found for a bearing of known length-to-diameter ratio if the eccentricity ratio is also known. Again evaluation of the load capacity, for example, also requires that the effective viscosity of the lubricant be known. The effective viscosity is, in turn, determined by the temperature rise which occurs in the lubricant as it flows through the bearing.

The dimensionless coefficients which describe the operation of a hydrodynamic journal bearing are, therefore, seen to be related to its operating characteristics as the eccentricity ratio adopted by the shaft is related to these dimensional characteristics. The proper design of a hydrodynamic bearing is, therefore, more complicated than a simple evaluation of a number of characteristics at a known value of eccentricity ratio. The proper evaluation of the effective viscosity of the lubricant, as determined by its associated temperature rise, is a crucial aspect of the design process. A similar situation was revealed in the study of plane thrust bearings (sub-section 5.5.4).

### 6.3.1   Dimensionless Coefficients for Design Purposes

As explained earlier, the dimensionless coefficients $\bar{W}$, $\bar{Q}$, $\bar{H}$ and $\bar{T}$ presented in sub-section 6.2.1 for a complete hydrodynamic journal bearing are functions of the length-to-diameter ratio, $b/d$, and the eccentricity ratio, $\varepsilon$. Although values of these coefficients have been presented in either tabular or graphical form, and may therefore be easily obtained from the previously cited references, it has been found that other forms of presentation are more suitable for design purposes. One such alternative presentation (Woolacot and Macrae, 1967) is outlined in the following derivations.

Initially put $K = (k/\rho C_p)$ in expression (6.2d) and rearrange it to get the following:

$$\frac{\eta N K}{\Delta T}\left(\frac{d}{C_d}\right)^2 = (1/\bar{T})$$

Multiply this expression by expression (6.2a), to obtain

$$\frac{WK}{d^2 \Delta T} = \frac{b}{d}(\bar{W}\bar{Q}/\bar{H})$$

Substitute for $\eta N$ from expression (6.2c) into expression (6.2d), to get

$$\frac{HK}{Nd^3 \Delta T}\left(\frac{d}{C_d}\right) = \frac{b}{d}(\bar{Q})$$

Finally, rearrange expression (6.2b), to obtain

$$\frac{Q}{Nd^3}\left(\frac{d}{C_d}\right) = \frac{b}{d}(\bar{Q})$$

The coefficients on the right-hand sides of the above expressions are functions of only the length-to-diameter ratio, $b/d$, and the operating eccentricity ratio, $\varepsilon$, for a particular bearing. These three coefficients are now redefined as follows:

$$\left.\begin{array}{l} \bar{W}_0 = \dfrac{b}{d}(\bar{W}\bar{Q}/\bar{H}) \\[2ex] \bar{T}_0 = (1/\bar{T}) \\[2ex] \bar{Q}_0 = \dfrac{b}{d}(\bar{Q}) \end{array}\right\} \qquad (6.3)$$

so that the rearranged dimensional groups may be expressed as

$$
\left.
\begin{aligned}
\frac{WK}{d^2 \Delta T} &= \bar{W}_0 \\[2mm]
\frac{\eta N K}{\Delta T}\left(\frac{d}{C_d}\right)^2 &= \bar{T}_0 \\[2mm]
\frac{Q}{Nd^3}\left(\frac{d}{C_d}\right) &= \bar{Q}_0 \\[2mm]
\frac{HK}{Nd^3 \Delta T}\left(\frac{d}{C_d}\right) &= \bar{Q}_0
\end{aligned}
\right\}
\tag{6.4}
$$

The values of $\bar{W}_0$, $\bar{T}_0$ and $\bar{Q}_0$ have been evaluated from knowledge of the appropriate values of $\bar{W}$, $\bar{Q}$ and $\bar{H}$ as functions of the length-to-diameter ratio, $b/d$, and the eccentricity ratio, $\varepsilon$. The important coefficients defined in (6.3), and shown on the right-hand sides of the expressions of (6.4), enable the full design of a hydrodynamic journal bearing to be approached in a routine manner, as shown in the examples which follow.

The values of the coefficients $\bar{W}_0$, $\bar{T}_0$ and $\bar{Q}_0$ are shown for practically useful ranges of the length-to-diameter ratio, $b/d$, and the eccentricity ratio, $\varepsilon$, in Figures 6.5, 6.6 and 6.7, respectively.

### 6.3.2 Example of Bearing Design with a Known Temperature Rise

A journal bearing must be designed to support a radial load of 1750 N which is placed on a shaft of 50 mm diameter. The shaft rotates at 10 000 rev/min and the minimum film thickness is $1.5 \times 10^{-2}$ mm. Experience shows that an eccentricity ratio of not less than 0.7 is appropriate and the desired temperature rise of the lubricant is to be 10°C. As the bearing will be well cooled, it may be assumed that only 0.7 of the heat generated will be convected from it by the oil. Design the bearing by determining its length, the diametral clearance and the type of oil required.

Assuming that the previously taken values of oil density and specific heat are applicable here (i.e. $\rho = 880 \text{ kg/m}^3$ and $C_p = 1880 \text{ J/kg·°C}$), then

$$
K = \frac{k}{\rho C_p} = \frac{0.7}{880 \times 1880} = 0.423 \times 10^{-6} \text{ m}^2 \cdot {}^\circ\text{C/N}
$$

To determine the length of bearing consider the first expression of (6.4):

$$
\bar{W}_0 = \frac{WK}{d^2 \Delta T} = \frac{1750(0.423 \times 10^{-6})}{(50 \times 10^{-3})^2 \times 10} = 29.6 \times 10^{-3}
$$

Reference to Figure 6.5 shows that this value of $\bar{W}_0$ corresponds to a number of different combinations of $b/d$ and $\varepsilon$. However, as it is prescribed that $\varepsilon$

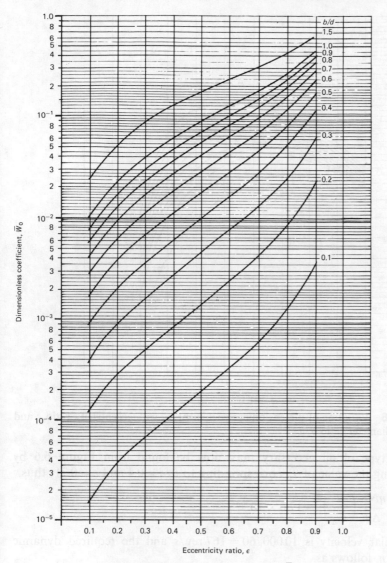

**Figure 6.5** Variation of dimensionless coefficient $\overline{W}_0$ with eccentricity ratio $\varepsilon$ and length-to-diameter ratio $b/d$

be at least 0.7, it is seen from this figure that a value of $\varepsilon = 0.71$ occurs when $b/d = 0.4$. It then follows that the length of the bearing is $0.4 \times 50 = 20$ mm.

The diametral clearance now follows from expression (6.1) by putting $\varepsilon = 0.71$ and $h = 1.5 \times 10^{-2}$ mm at $\theta = \pi$; thus,

$$C_d = \frac{2(1.5 \times 10^{-2})}{1 - 0.71} = 0.103 \text{ mm}$$

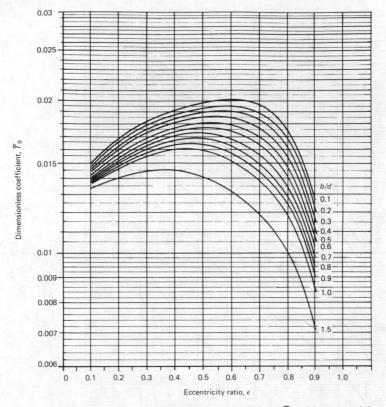

**Figure 6.6** Variation of dimensionless coefficient $\bar{T}_0$ with eccentricity ratio $\varepsilon$ and length-to-diameter ratio $b/d$

The type of oil necessary may now be found from Figure 6.6 by determining the value of $\bar{T}_0$ corresponding to $b/d = 0.4$ and $\varepsilon = 0.71$; thus,

$$\bar{T}_0 = \frac{\eta NK}{\Delta T}\left(\frac{d}{C_d}\right)^2 = 17.6 \times 10^{-3}$$

The angular velocity is $10\,000/60 = 167\,\text{rev/s}$ and the required dynamic viscosity, $\eta$, follows as

$$\eta = 17.6 \times 10^{-3}\left\{\frac{\Delta T}{NK}\left(\frac{C_d}{d}\right)^2\right\}$$

$$= 17.6 \times 10^{-3}\left\{\frac{10}{167 \times 0.423 \times 10^{-6}}\left(\frac{0.103 \times 10^{-3}}{50 \times 10^{-3}}\right)^2\right\}$$

$$\eta = 10.5 \times 10^{-3}\,\text{kg/m}\cdot\text{s} = 10.5\,\text{cP}$$

The corresponding kinematic viscosity is $10.5/0.88 = 11.9\,\text{cSt}$ and reference to Figure 5.2 shows that Tellus 22 oil has this viscosity at a

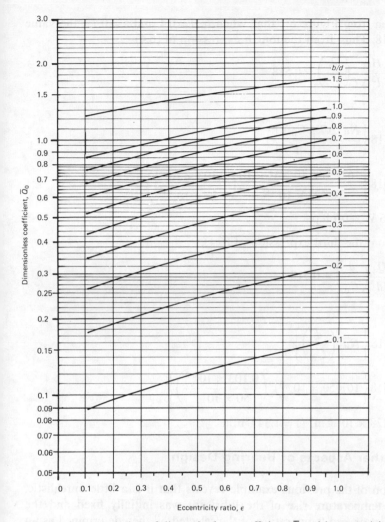

**Figure 6.7** Variation of dimensionless coefficient $\bar{Q}_0$ with eccentricity ratio $\varepsilon$ and length-to-diameter ratio $b/d$

temperature of 59 °C. As the temperature rise is 10 °C, the inlet or feed temperature of the oil must be $59 - 10 = 49$ °C. The kinematic viscosity of the oil at 49 °C is seen to be 16.2 cSt, so that the dynamic viscosity of the oil as it enters the bearing, $\eta_f$, is $16.2 \times 0.88 = 14.3$ cP.

Thus, the length of the bearing has been determined as 20 mm, the diametral clearance has been determined as 0.103 mm, and a suitable oil has been identified.

Two other parameters of interest may also be determined – the power required to rotate the shaft and the flow rate required to ensure proper operation of the bearing.

It is seen from Figure 6.7 that when $b/d = 0.4$ and $\varepsilon = 0.71$, the value of $\bar{Q}_0$ is 0.518. Thus, from the fourth expression of (6.4),

$$\bar{Q}_0 = \frac{Hk}{Nd^3 \Delta T}\left(\frac{d}{C_d}\right) = 0.518$$

or

$$H = 0.518\left\{\frac{Nd^3 \Delta T}{K}\left(\frac{C_d}{d}\right)\right\}$$

$$= 0.518\left\{\frac{167(50 \times 10^{-3})^3 \times 10}{0.423 \times 10^{-6}}\left(\frac{0.103 \times 10^{-3}}{50 \times 10^{-3}}\right)\right\}$$

$$H = 526\ \mathrm{W} = 0.7\ \mathrm{h.p.}$$

Similarly, from the third expression of (6.4),

$$\bar{Q}_0 = \frac{Q}{Nd^3}\left(\frac{d}{C_d}\right) = 0.518$$

or

$$Q = 0.518\left\{Nd^3\left(\frac{d}{C_d}\right)\right\}$$

$$= 0.518\left\{167(50 \times 10^{-6})^3\left(\frac{0.103 \times 10^{-3}}{50 \times 10^{-3}}\right)\right\}$$

$$Q = 22.28 \times 10^{-6}\ \mathrm{m^3/s} = 1.34\ \mathrm{l/min}$$

### 6.3.3 Further Aspects of Bearing Design

Consideration of the previous example reveals that it is a little unrealistic, in that the temperature rise of the lubricant was initially fixed and the necessary inlet or feed temperature of the selected oil then determined as an operational requirement. However, in most practical cases it is the inlet temperature of the oil that is known, perhaps approximately, so that the temperature rise must be determined. A similar situation was considered in the case of thrust bearings in sub-section 5.5.4, where an iterative procedure was used to determine the initially unknown temperature rise as one of a number of interrelated variables.

The application of a similar procedure to the design of hydrodynamic journal bearings is described in a later example (sub-section 6.5.1).

The flow rate of fluid necessary to ensure proper operation of the journal bearing considered in the last example was easily determined. However, further consideration must be given to the associated problem of specifying the details of the supply groove in order that the required flow rate of fluid be able to enter the hydrodynamically generated film. Common grooving arrangements are discussed in the next section.

## 6.4   OIL-SUPPLY GROOVES

In sub-section 6.1.3 the general shape and position of an oil-supply groove in a bearing was shown schematically in Figure 6.4. In most cases the actual groove is machined in the cylindrical liner which constitutes the bearing proper and provides the stationary surface of the clearance. The type of groove shown in this figure is known as the rectangular feed groove, but several other widely used geometries exist (Martin, 1983a). Some of the more commonly encountered grooving arrangements are discussed in the next sub-section.

### 6.4.1   Geometry of Oil-supply Grooves

The purpose of an oil-supply groove is to supply the necessary quantity of oil to the start of the convergent section of the clearance – i.e. to section $E$, where $\theta$ is zero, in Figure 6.3. A full specification of an oil-supply groove demands a description of its basic geometric shape, its dimensions and its position in the journal bearing relative to the line of action of the load. In this sub-section it will be assumed that the direction of the load does not vary, but some of the grooving arrangements discussed here are suitable for loads which vary in direction or for shafts which are able to rotate in both directions.

In all cases it is necessary for the lubricant to be pumped by external means to the supply groove, and the determination of the supply pressure, $p_f$, which is to be maintained in the supply groove is part of the design procedure.

Four basic geometries of oil-supply grooves are shown in Figure 6.8 and discussed below.

The provision of a single circular hole in a bearing, as shown in Figure 6.8(a), represents the simplest form of grooving. The presence of a complete circumferential groove of the type shown in Figure 6.8(b) effectively provides two separate bearings which are able to carry loads applied in any direction and to accept shaft rotation in both directions. A variation of this arrangement is shown in Figure 6.8(c), where the circumferential groove extends over about 180°. This arrangement allows loads to be applied over a limited range of direction but still accepts shaft rotation in either direction. Figure 6.8(d) illustrates the general arrangement of a rectangular groove in its basic form, but variations occur because of the choice of position available.

With reference to Figure 6.2, it would seem possible to locate a rectangular oil supply groove in any position between sections $G$ and $E$ in

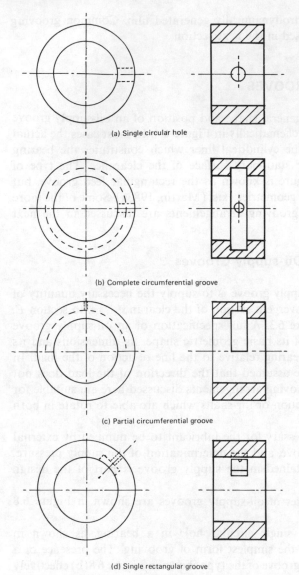

(a) Single circular hole

(b) Complete circumferential groove

(c) Partial circumferential groove

(d) Single rectangular groove

**Figure 6.8** Types of lubricant-supply grooves

the divergent portion of the clearance. However, the practically acceptable positions correspond in most cases to those shown in Figure 6.9.

Two common locations for a single rectangular oil-supply groove are at 45° to the load line, as shown in Figure 6.9(a), and at 90° to the load line, as shown in Figure 6.9(b). Naturally, the direction in which these angles are measured from the load line must correspond to the direction of shaft rotation. It follows, therefore, that bearings equipped with only a single

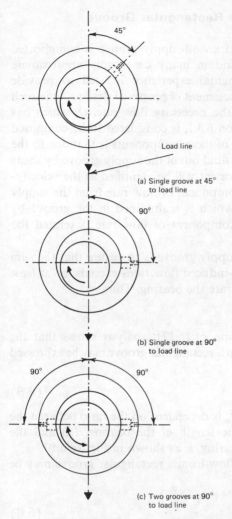

(a) Single groove at 45°
to load line

(b) Single groove at 90°
to load line

(c) Two grooves at 90°
to load line

**Figure 6.9** Positions of rectangular lubricant-supply grooves

rectangular groove are not suitable for shafts which are able to rotate in both directions or for loads which vary significantly in direction.

Provision of two rectangular grooves at right angles to the load line overcomes these limitations to some extent, as shown in Figure 6.9(c). The arrangement of Figure 6.9(c) is commonly adopted for constructional reasons, as bearings are often split along a line at right angles to the load line in manufacture.

Further consideration of oil-supply grooves will be limited to a discussion of the flow of fluid from and the design of a single rectangular groove in the locations shown in Figures 6.9(a) and (b).

### 6.4.2  Flow Rate from a Single Rectangular Groove

The evaluation of the flow rate from an oil-supply groove is complicated (Cameron, 1981; Martin, 1983a), and in many cases only approximate analyses are available and experimental experience is used to provide adequately accurate means for the assessment of proposed designs. One such method is described here, in which the necessary flow rate, $Q$, which has already been established in sub-section 6.3.1, is considered to be composed of two distinct components. The first of these components is that due to the effect of the moving shaft in dragging fluid out of the supply groove by shear action (sub-section 5.3.1). This component will be identified as the velocity-induced flow rate, $Q_u$. The second component of flow rate from the supply groove is that due to the pressure which is maintained in the groove by external means, e.g. a pump. This component of flow rate is termed the pressure-induced flow rate, $Q_p$.

The object of designing an oil-supply groove is to ensure that the sum of the velocity-induced and pressure-induced flow rates exceeds, or at least equals, the flow rate required to operate the bearing. Thus,

$$Q_u + Q_p \geqslant Q$$

In parallel with the second expression of (5.12), analysis shows that the velocity-induced flow rate of fluid from a rectangular groove may be expressed as

$$Q_u = \bar{Q}_u \{ N b d C_d \} \tag{6.5}$$

where the dimensionless coefficient $\bar{Q}_u$ is dependent on the ratio between the length of the supply groove and the length of the bearing, $l/b$ and the operating eccentricity ratio of the bearing, $\varepsilon$, as shown in Figure 6.10.

Similarly, the pressure-induced flow from a rectangular groove may be given as follows:

$$Q_p = \bar{Q}_p \left\{ \frac{p_f C_d^3}{\eta_f} \right\} \tag{6.6}$$

Many assumptions have been made in the analyses which are summarised by expressions (6.5) and (6.6), as the geometry of a rectangular groove results in complicated flow patterns. Simplifications have made the curves of Figure 6.10 independent of the length-to-diameter ratio of the bearing, $b/d$, and the angular position of the groove, but the length of the groove, $l$, does, of course, appear in the ratio $l/b$.

The value of the dimensionless coefficient $\bar{Q}_p$ is conveniently expressed as the product of two other coefficients, $\bar{Q}_1$ and $\bar{Q}_2$, which are given by the curves of Figures 6.11(a) and 6.11(b), respectively. Thus, $\bar{Q}_p = \bar{Q}_1 \bar{Q}_2$, so that

$$Q_p = \bar{Q}_1 \bar{Q}_2 \left\{ \frac{p_f C_d^3}{\eta_f} \right\} \tag{6.7}$$

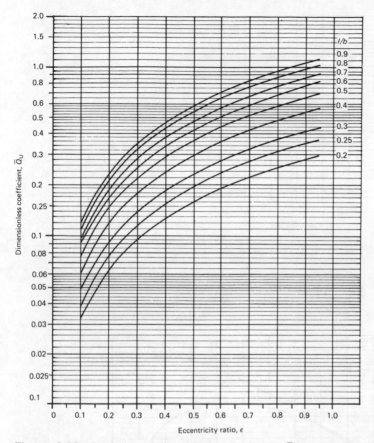

**Figure 6.10**    Variation of dimensionless coefficient $\bar{Q}_u$ with eccentricity ratio $\varepsilon$ and length ratio $l/b$

The value of $\bar{Q}_1$ is affected by the dimensions of the rectangular groove, $a$ and $l$, as they are introduced through the ratios $a/b$ and $l/b$, while the two selected angular positions of the groove and the eccentricity ratio are reflected in the value of $\bar{Q}_2$.

Acceptance of the above presentation now enables the dimensions of the supply groove required to ensure operation of the bearing of sub-section 6.3.2 to be determined along with the necessary supply pressure, as shown in sub-section 6.4.4.

## 6.4.3    Effective Temperature and Viscosity in Hydrodynamic Journal Bearings

It was explained in sub-section 5.5.1 that account is usually taken of the decrease in the viscosity of oil as it passes through a thrust bearing by basing calculations on an effective viscosity. The effective viscosity was defined for

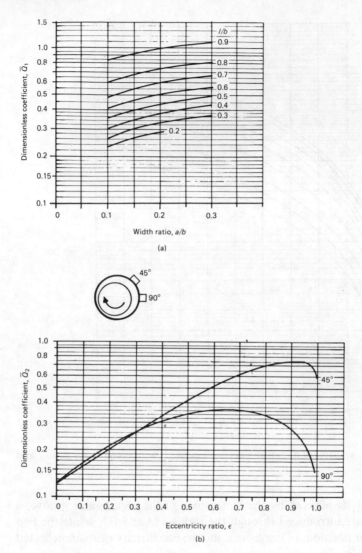

**Figure 6.11** Variation of dimensionless coefficients $\bar{Q}_1$ and $\bar{Q}_2$ with eccentricity ratio $\varepsilon$, length ratios $a/b$ and $l/b$, and groove position

a thrust bearing as that at the effective or mean temperature of the oil during its passage through the bearing. Most of the oil which enters a thrust bearing passes completely through the converging clearance and is nearly all available to convect heat from the bearing. Consequently, the effective temperature was defined as the sum of the inlet temperature and a *half* of the temperature rise as given by the power balance (sub-section 5.5.4).

However, in most journal bearings the width of the load-bearing film is much less than its length. For example, in a journal bearing with the relatively high length-to-diameter ratio of unity the axial width of the load-bearing film is only about $\frac{2}{3}$ of its circumferential length, as it extends over an arc of only about 180°C. Therefore, a considerable proportion of the oil which enters the convergent, and load-bearing, clearance escapes axially, so that not all of the oil which enters the bearing is equally effective in convecting away the heat which is generated in the load-bearing film.

It follows that the temperature of the oil at exit from the load-bearing arc (section $G$ in Figure 6.3) attains a higher value than that which would have occurred if all of the oil which entered at section $E$ had passed completely through the bearing.

This higher exit temperature results in the effective temperature of the oil in a journal bearing being more accurately represented by the sum of the inlet temperature and the *whole* of the temperature rise found from the power balance rather than a half of it, as in the case of a thrust bearing. This way of assigning the value of the effective temperature was used in sub-section 6.3.2 and will be used later in another example (sub-section 6.5.1)

### 6.4.4   Example of Determination of the Dimensions of a Supply Groove

This example simply completes the design of the bearing commenced in sub-section 6.3.2 by determining the dimensions of the rectangular oil-supply groove required, subject to the condition that the supply pressure in the groove is limited to a maximum of 0.4 N/mm$^2$.

It has already been shown in sub-section 6.3.2 that the flow rate required to operate the bearing is 1.34 l/min and that the viscosity of the oil in the supply groove is 14.3 cP. It is required that $Q_u + Q_p \geqslant Q$.

Study of Figure 6.10 reveals that an estimate of the length of the groove, $l$, must be made at this stage by choosing a value for the ratio $l/b$. Take the geometrically reasonable value of $l/b = 0.65$ and $\varepsilon = 0.71$ in Figure 6.10, to get $\bar{Q}_u = 0.7$.

The velocity-induced flow rate, $Q_u$, then follows from Equation (6.5) as

$$Q_u = \bar{Q}_u(NbdC_d)$$

$$= 0.7\{167(20 \times 10^{-3})(50 \times 10^{-3})(0.103 \times 10^{-3})\}$$

$$Q_u = 12 \times 10^{-6}\,\text{m}^3/\text{s} = 0.72\,\text{l/min}$$

The pressure-induced flow, $Q_p$, may now be determined by use of expression (6.7) and Figures 6.11(a) and 6.11(b). From Figure 6.11(a), with the assumption that the ratio $a/b$ is 0.25 and $l/b = 0.65$, the value of $\bar{Q}_1$ is seen to be 0.59. If it is then decided to place the supply groove at an angle of 45° to the load line, the value of $\bar{Q}_2$ is given in Figure 6.11(b) as 0.62 for an eccentricity ratio of 0.71.

The pressure-induced flow rate then follows from Equation (6.7) as

$$Q_p = \bar{Q}_1 \bar{Q}_2 \left( \frac{p_f C_d^3}{\eta_f} \right)$$

$$= 0.59 \times 0.62 \left\{ \frac{(0.4 \times 10^6)(0.103 \times 10^{-3})^3}{14.3 \times 10^{-3}} \right\}$$

$$Q_p = 11.2 \times 10^{-6} \, \text{m}^3/\text{s} = 0.67 \, \text{l/min}$$

Thus,

$$Q_u + Q_p = 0.72 + 0.67 = 1.39 \, \text{l/min}$$

This flow rate is greater than the required value of 1.34 l/min and is therefore acceptable.

The length of the axial groove is therefore

$$l = 20 \times 0.65 = 13 \, \text{mm}$$

and its width is

$$a = 20 \times 0.25 = 5 \, \text{mm}$$

## 6.5  DETERMINATION OF TEMPERATURE RISE

The form of the dimensionless coefficients introduced in sub-section 6.3.1 provide a means by which the temperature rise in a journal bearing, relative to a given inlet temperature, may be determined, along with appropriate values of length-to-diameter ratio, eccentricity ratio and diametral clearance. Such evaluations inevitably involve the viscosity temperature characteristics of oils and these, too, may be included directly in the design process.

Iteration is usually necessary to achieve a satisfactory design, and one such scheme of iteration is illustrated in Figure 6.12 and described briefly below.

In general, the diameter of the bearing and the load applied to it are known, so that $\bar{W}_0$ may be evaluated for an *initially* assumed value of the temperature rise, $\Delta T$, from the first expression of (6.4). This value of $\bar{W}_0$ may be entered in Figure 6.5 and corresponding values of $b/d$ and $\varepsilon$ selected. Once $\varepsilon$ is known, the diametral clearance, $C_d$, may be calculated, as the minimum film thickness is usually specified. Figure 6.6 may now be entered with the known values of $b/d$ and $\varepsilon$, to determine the value of $\bar{T}_0$. The second expression of (6.4) then enables the necessary value of the effective dynamic viscosity to be calculated.

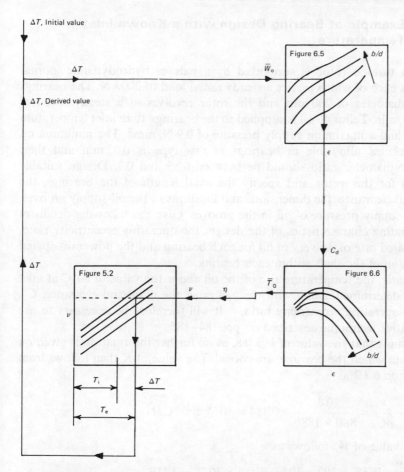

**Figure 6.12** Iterative scheme for design of hydrodynamic journal bearings

The corresponding value of the kinematic viscosity, $v$, is then entered in Figure 5.2, along with the now known effective temperature of the oil, as given by the sum of the inlet temperature and the assumed value of temperature rise, $\Delta T$.

At this stage the necessary kinematic viscosity should correspond with the characteristic of the selected oil at the calculated effective temperature. If no correspondence exists, then the process may be repeated by choosing another value of the temperature rise, $\Delta T$, until the variables involved give consistency in all three of the characteristics represented in Figure 6.12.

Once this agreement has been established, the other important operating parameters of the bearing may be studied. The following example illustrates this method of iteration for determining a consistent set of variables.

### 6.5.1  Example of Bearing Design with a Known Inlet Temperature

A steam turbine rotor is supported by a pair of hydrodynamic journal bearings, each of which carries a steady radial load of 200 kN. The bearings have a diameter of 200 mm and the rotor revolves at a steady speed of 3000 rev/min. Tellus 46 oil is supplied to the bearings at an inlet temperature of 50°C and a maximum supply pressure of 0.9 N/mm². The minimum oil film thickness allowable in bearings of this type is 0.02 mm and their length-to-diameter ratio should be between 0.55 and 0.7. Design suitable bearings for this rotor and specify the axial length of the bearings, the diametral clearance, the dimensions and locations of the oil-supply grooves, and the supply pressure of oil in the grooves. Give the following details of the operating characteristics of the design: the operating eccentricity ratio, the required rate of delivery of oil for each bearing and the power dissipated by rotation of the shaft within each bearing.

Clearly, the temperature rise of the oil above the value of 50°C at inlet is to be determined, along with the $b/d$ ratio, the diametral clearance, $C_d$, and the operating eccentricity ratio, $\varepsilon$. It will therefore be necessary to use the iterative technique described on pp. 184–185.

Assume that the value of $k$ is 0.8, as no further information is given on the way in which the bearings are cooled. The value of $K$ then follows from sub-section 6.3.2 as

$$K = \frac{k}{\rho C_p} = \frac{0.8}{880 \times 1880} = 0.484 \times 10^{-6}\,\mathrm{m^2 \cdot {}^\circ C/N}$$

and the value of $\bar{W}_0$ follows as

$$\bar{W}_0 = \frac{WK}{d^2 \Delta T} = \frac{200 \times 10^3 \times 0.484 \times 10^{-6}}{(200 \times 10^{-3})^2 \Delta T} = \frac{2.418}{\Delta T} \tag{a}$$

The required diametral clearance is, from expression (6.1),

$$C_d = \frac{2(0.02 \times 10^{-3})}{1 - \varepsilon} = \frac{0.04 \times 10^{-3}}{1 - \varepsilon}\,\mathrm{m} \tag{b}$$

and $\bar{T}_0$ is, as $N = 3000/60 = 50$ rev/min:

$$\bar{T}_0 = \frac{\eta N K}{\Delta T}\left(\frac{d}{C_d}\right)^2 = \eta\,\frac{50 \times 0.484 \times 10^{-6}}{\Delta T}\left(\frac{200 \times 10^{-3}}{C_d}\right)^2$$

or

$$\bar{T}_0 = 0.967 \times 10^{-6}\left(\frac{\eta}{\Delta T C_d^2}\right) \tag{c}$$

In addition, the effective temperature of the oil will be $(50 + \Delta T)°C$. The sequence of calculations outlined in Section 6.5 is now carried out for various values of $\Delta T$ until convergence occurs between Figure 6.5 and Figure 6.6 and the viscosity temperature characteristics for Tellus 46 oil given in Figure 5.2. Take $\Delta T = 20°C$; then, from expression (a) above,

$$\bar{W}_0 = \frac{2.418}{20} = 0.129$$

Reference to Figure 6.5 shows that this value corresponds to $b/d = 0.7$ and $\varepsilon = 0.76$ or $b/d = 0.5$ and $\varepsilon = 0.81$. Take the first pair of values and proceed to find $C_d$ from (b) as

$$C_d = \frac{0.04 \times 10^{-3}}{1 - 0.76} = 0.167 \times 10^{-3} \, m$$

Put this value in (c) and recall that, as both $b/d$ and $\varepsilon$ are now known, the value of $\bar{T}_0$ may be read off Figure 6.6 as $14.52 \times 10^{-3}$; hence,

$$14.52 \times 10^{-3} = 0.967 \times 10^{-6} \left\{ \frac{\eta}{20(0.167 \times 10^{-3})^2} \right\}$$

from which $\eta = 0.0084 \, kg/m \cdot s$ or $\nu = 9.55 \, cSt$, a value which does not correspond with the viscosity of 15 cSt given for Tellus 46 at an effective temperature of $(50 + 20) = 70°C$ in Figure 5.2.

Repetition of this process for different values of the temperature rise, $\Delta T$, in accordance with the above specimen calculation and the method of Section 6.5 yields the following set of consistent results for $\Delta T = 25°C$.

From (a)

$$\bar{W}_0 = \frac{2.418}{25} = 0.0967$$

which corresponds to $b/d = 0.6$ and $\varepsilon = 0.77$ in Figure 6.5.

The value of $C_d$ then follows from (b):

$$C_d = \frac{0.04 \times 10^{-3}}{1 - 0.77} = 0.174 \times 10^{-3} \, m$$

With $b/d = 0.6$ and $\varepsilon = 0.77$, the value of $\bar{T}_0$ may be read off Figure 6.6 as $15 \times 10^{-3}$, so that (c) becomes

$$15 \times 10^{-3} = 0.967 \times 10^{-6} \left\{ \frac{\eta}{25(0.174 \times 10^{-3})^2} \right\}$$

for which $\eta = 0.0117 \, kg/m \cdot s$ or 13.3 cSt.

The effective temperature of the oil is $(50 + 25) = 75°C$ and Figure 5.2 shows that the kinematic viscosity of Tellus 46 at 75°C is about 13.5 cSt, so that the values corresponding to a temperature rise of 25°C are consistent.

With $b/d = 0.6$ and $\varepsilon = 0.77$, the value of $\bar{Q}_0$ may be read off Figure 6.7 as 0.78, so that the power loss follows from the fourth expression of (6.4) as

$$H = \bar{Q}_0 \left\{ \frac{Nd^3 \Delta T}{K} \left( \frac{C_d}{d} \right) \right\}$$

$$= 0.78 \left\{ \frac{50(200 \times 10^{-3})^3 25}{0.484 \times 10^{-6}} \left( \frac{0.174 \times 10^{-3}}{200 \times 10^{-3}} \right) \right\}$$

$$H = 14 \times 10^3 \text{ W} = 18.8 \text{ h.p.}$$

Again the required flow rate follows from the third expression of (6.4) as

$$Q = \bar{Q}_0 \left\{ Nd^3 \left( \frac{C_d}{d} \right) \right\}$$

$$= 0.78 \left\{ 50(200 \times 10^{-3})^3 \left( \frac{0.174 \times 10^{-3}}{200 \times 10^{-3}} \right) \right\}$$

$$Q = 0.271 \times 10^{-3} \text{ m}^3/\text{s} = 16.28 \text{ l/min}$$

The velocity-induced flow, $Q_u$, may be found from Figure 6.10 with an initially assumed value of $l/b$ as, say, 0.6, so that $\bar{Q}_u$ becomes 0.7 and $Q_u$ is then

$$Q_u = \bar{Q}_u(NbdC_d)$$

$$= 0.7\{50(0.6 \times 200 \times 10^{-3})(200 \times 10^{-3})(0.174 \times 10^{-3})\}$$

$$Q_u = 0.146 \times 10^{-3} \text{ m}^3/\text{s} = 8.76 \text{ l/min}$$

Thus, the minimum pressure-induced flow rate is $Q - Q_u = 16.28 - 8.76 = 7.52$ l/min $= 0.125 \times 10^{-3}$ m$^3$/s.

Now assume that the supply groove is placed at 90° to the load line, with the ratio $a/b$ set at 0.25. Figures 6.11 then give $\bar{Q}_1 = 0.54$ and $\bar{Q}_2 = 0.35$, so that the pressure-induced flow becomes, from expression (6.7),

$$Q_p = \bar{Q}_1 \bar{Q}_2 \left\{ \frac{p_f C_d^3}{\eta_f} \right\}$$

Now the viscosity of the oil at an inlet temperature of 50°C is, from Figure 5.2, 30 cSt or $30(0.88)/1000 = 0.0263$ kg/m·s. The supply pressure required to provide the necessary flow rate of oil may now be found from the above expression as

$$0.125 \times 10^{-3} = 0.54 \times 0.35 \left\{ \frac{p_f(0.174 \times 10^{-3})^3}{0.0263} \right\}$$

$$p_f = 3.31 \times 10^6 \text{ N/m}^2 = 3.31 \text{ N/mm}^2$$

This value of $p_f$ is in excess of the maximum available supply pressure of 0.9 N/mm$^2$. Changes in the various assumptions may now be made in order to evolve an acceptable design. By taking $l/b = 0.8$ it is found that $\bar{Q}_u$ becomes 0.9 and $Q_u = 0.187 \times 10^{-3}$ m$^3$/s. The required pressure-induced flow then becomes

$$Q_p = 0.271 \times 10^{-3} - 0.187 \times 10^{-3} = 0.084 \times 10^{-3} \, \text{m}^3/\text{s}$$

With the supply groove placed at 45° to the load line and $a/b$ maintained at 0.25, it is found that $\bar{Q}_1 = 0.78$ and $\bar{Q}_2 = 0.67$; thus,

$$0.084 \times 10^{-3} = 0.78 \times 0.67 \left\{ \frac{p_f (0.174 \times 10^{-3})^3}{0.0263} \right\}$$

$$p_f = 0.8 \times 10^6 \, \text{N/m}^2 = 0.8 \, \text{N/mm}^2$$

This value of $p_f$ is less than 0.9 and therefore acceptable. The length of the bearing is $0.6(200) = 120$ mm and the dimensions of the oil-supply groove are as follows:

$$l = 0.8(120) = 96 \text{ mm}$$

$$a = 0.25(120) = 30 \text{ mm}$$

## 6.6 INSTABILITY IN JOURNAL BEARINGS

Rotating shafts in hydrodynamic journal bearings are subject to two possible forms of instability.

The first, often called synchronous whirl, is caused by a periodic disturbance outside the bearing, such as inbalance of the rotor, which excites the bearing–shaft system into resonance. Since it is impossible to produce a perfectly balanced shaft, the centre of the shaft will not normally be stationary but will describe a tiny circle or other closed orbit around the equilibrium position. If this orbit is constant, the whirl is stable. However, if the speed of rotation is progressively increased, at some stage the rotational frequency of the rotor inbalance will coincide with the first natural frequency of the rotor system and the whirl orbit will increase. At this speed either the whirl will become stable at a bigger amplitude or failure will take place. The magnitude of this first critical speed depends on the rotor stiffness and inertia and on the stiffness and damping characteristics of the bearings. On further increasing the speed, the whirl subsides until the second critical speed is reached and resonance will once more occur. If the shaft is accelerated rapidly, these critical speeds can be passed through before the whirl orbit has time to grow large enough to damage the bearing. Clearly, it is desirable that the running speed be selected to avoid these critical speeds.

The other form of instability is called 'half-speed whirl' or 'oil-whip' and is induced by the lubricant film itself (Holmes, 1963). We have seen that, when loaded vertically downward, the centre of the journal does not simply move down in the direction of the load but also moves around the bearing in the direction of rotation (Figure 6.3). In some cases this motion can continue, so that the shaft centre describes a circular orbit. If this rotation takes place at half the rotational speed, it will coincide with the mean rotational speed of the lubricant and no relative motion between the film shape and the lubricant will occur, so that the hydrodynamic mechanism will be destroyed.

This phenomenon will also take place at a certain threshold speed but, unlike synchronous whirl, there is no possibility of running through half-speed whirl, further increases in speed only accelerating the failure of the bearing.

### 6.6.1 Prediction and Suppression of Instability

Synchronous whirl is usually avoided by either increasing the bearing stiffness, so that the first critical speed is well above the commonly used running speed of the rotor, or decreasing the stiffness, so that the critical speeds are quickly passed through and normal running takes place where the attenuation is large.

Another means of suppressing or allowing for whirl is the introduction of extra damping into the system. This may be done by flexibly mounting the bearings in, for example, rubber or metal diaphragms.

Half-speed whirl is most likely to become a problem in lightly loaded bearings and it is often possible to raise the threshold speed by redesigning the bearing to run at a higher eccentricity ratio. It is, therefore, very troublesome in vertically mounted spindles, where the horizontal load is negligible and the operating eccentricity ratio correspondingly small.

By mounting the bearings flexibly, as described earlier, extra damping may be incorporated into the system and some of the energy generated by the whirling will be absorbed in the damping medium. This form of instability is very common in bearings lubricated by a gas, because of the very poor damping properties of the lubricant itself. Indeed, this is often the limitation on the use of gas journal bearings, rather than the lack of load capacity (Marsh, 1964).

The criterion given for the onset of whirl (ESDU 84031, 1984) can be written as $McN^2/W < 0.2$, where $c$ is the radial clearance, $M$ the mass of the rotating shaft, $W$ the load carried by the bearing and $N$ the angular velocity. For a bearing with a purely gravitational loading due to the mass of the shaft, this becomes $cN^2/g < 0.2$. More information on bearing instabilities may be found in the references (ESDU 77013, 1977).

An effective way of preventing or delaying half-speed whirl is to interfere with the circumferential symmetry of the bearing. An axial groove in the low-pressure region, two or more lobes in the bore or even accidental out-of-roundness errors in manufacture can be sufficient to delay the onset of whirl (Malik et al., 1982; Lanes et al., 1982).

## 6.7 PRACTICAL DESIGN OF JOURNAL BEARINGS

This chapter has described the operation of hydrodynamic journal bearings and presented a method of evolving complete designs of bearings for relatively uncomplicated cases. Other methods of bearing design have been derived (Barwell, 1979; ESDU 84031, 1984) and methods of optimisation proposed (Moes and Bosma, 1971). Many considerations enter into the full design of journal bearings, and the choice of materials to be used in their construction is of importance (ESDU 88018, 1988). Special considerations must be taken into account in some applications where, for example, dynamic loading is of significance (Martin, 1983b).

## 6.8 PROBLEMS

1  For a range of eccentricity ratio $\varepsilon$ between 0.2 and 0.7, the dimensionless load coefficient, $\bar{W}$ (defined in expression 6.2a), is found to be related to the eccentricity ratio by the following expressions:

$$\log_{10}(\bar{W}) = 2.2496\varepsilon - 1.194, \text{ when } b/d = 0.3$$

$$\log_{10}(\bar{W}) = 1.918\varepsilon - 0.347, \text{ when } b/d = 0.8$$

A complete journal bearing has a diameter of 80 mm, a length of 64 mm and a diametral clearance of 0.08 mm, uses an oil of effective dynamic viscosity 0.04 kg/m·s and supports a shaft which rotates at 600 rev/min.

(a) The radial load imposed on the bearing in service varies between 3000 N and 18 000 N. Estimate the corresponding values of eccentricity ratio.
    [0.267 and 0.673]

(b) The bearing is now modified by machining in it a complete centrally located circumferential groove of width 16 mm, as shown in Figure 6.8(b). Find the value of the diametral clearance required to ensure that this bearing operates at an eccentricity ratio of 0.7 under the full load of 18 000 N.
    [0.036 mm]

2  It is found that in the range of eccentricity ratio $\varepsilon$ between 0.2 and 0.7 the dimensionless load coefficient, $\bar{W}$ (defined in expression 6.2a), may be related to the eccentricity ratio and length-to-diameter ratio of the bearing, $b/d$, by the following expression:

$$\bar{W} = 0.68\left(\frac{b}{d}\right)^{1.76} e^{4.2\varepsilon}$$

A journal bearing has a diameter of 80 mm, a length of 64 mm and a diametral clearance of 0.08 mm, uses an oil of effective dynamic viscosity 0.04 kg/m·s and operates at 600 rev/min.

(a) The load carried by the bearing in service varies between 3000 N and 15 000 N. Estimate the corresponding range of eccentricity ratio and the minimum film thickness under the larger load.
    [0.276–0.66; 0.014 mm]
(b) What should be the length of the bearing if it is to operate at an eccentricity ratio of 0.5 under a load of 3000 N? Its other dimensions, the viscosity of the oil and the speed of rotation remain unchanged.
    [45.6 mm]
(c) The length of the bearing is now fixed at 48 mm while carrying a load of 3000 N. Investigate the variation of eccentricity ratio and minimum film thickness when the diametral clearance varies between 0.050 mm and 0.070 mm. Hence select the diametral clearance which gives a minimum film thickness of 0.02 mm.
    [0.058 mm with $\varepsilon = 0.312$]

3   A short rotor of total mass 1000 kg is supported by two identical hydrodynamic journal bearings of length 50 mm, diameter 100 mm and diametral clearance 0.14 mm. The rotor is driven at a steady speed of 1200 rev/min and the effective dynamic viscosity of the oil is 30 cP. It may be assumed that for a journal bearing with a length-to-diameter ratio of 0.5 the load coefficient, $\bar{W}$ (defined in expression 6.2a), is related to the eccentricity ratio, $\varepsilon$, by the following expression:

$$\bar{W} = 0.18e^{4.8\varepsilon}$$

(a) Show that the *stiffness* of the bearing for small displacements about any steady operating eccentric position defined by the eccentricity, $e$, may be found as:

$$\frac{\partial W}{\partial e} = \frac{\eta N b d}{C_d}\left(\frac{d}{C_d}\right)^2 (9 \cdot 6\bar{W})$$

where $\partial W/\partial e$ is the stiffness, or rate of change of load with small displacement, $\eta$ is the dynamic viscosity of the lubricant, $N$ is the rotational speed of the rotor, $b$ is the length of the bearing, $d$ is its diameter and $C_d$ is the diametral clearance.
(b) Show that the bearings of the assembly described above operate with a steady operating eccentricity ratio of about 0.6.
(c) Estimate the stiffness of each bearing and the fundamental undamped free frequency of vibration of the rotor in its bearings.
    [$335 \times 10^6$ N/m, 130 c/s]

**4** A circular shaft rotates at 2400 rev/min and is supported in two hydrodynamic journal bearings, each of which carries a steady radial load of 60 kN. The shaft has a diameter of 240 mm and one of the Tellus range of oils may be supplied to the bearings at an inlet temperature of 30°C and a maximum supply pressure of 0.8 N/mm². Experience shows that the minimum film thickness in the bearings must not be less than 0.04 mm. Each bearing is to have an axial length of 96 mm, and as the bearings' housing is very well insulated, it is estimated that 90% of the heat generated by rotation of the shaft is convected away by the oil. Design a suitable bearing for this shaft and give the following details of the full design:

(a) The operating eccentricity ratio in the bearings.
(b) The diametral clearance between the shaft and bearings.
(c) The dimensions of the oil-supply groove and the pressure in it.
(d) The required rate of delivery of oil to each bearing.
(e) The mean temperature rise of the oil as it flows through the bearings.

[One consistent set of answers is as follows:
(a) 0.7, (b) 0.27 mm, (c) 77 mm × 24 mm, (d) 19.2 l/min, (e) 20.3 °C]

# Chapter 7
# The Lubrication of Highly Loaded Contacts

## 7.1 INTRODUCTION

In Chapter 5 the hydrodynamic build-up of a lubricating film between the moving surfaces was examined. However, if the pressure between the surfaces becomes very high, there are two extra effects which have to be taken into account when assessing the performance of the lubricated contact: the increase of viscosity with pressure and the local elastic deformation of the surfaces. Both these effects lead to an increase in film thickness compared with the simple hydrodynamic theory, and the mechanism of film generation in this case is known as elastohydrodynamic lubrication (EHL).

This situation is most likely to occur either where the load is concentrated over a small area, owing to the surfaces being non-conforming, such as in rolling element bearings, gears and cams, or where at least one of the surfaces is of a material of low elastic modulus, such as a rubber tyre rolling on a road.

## 7.2 EQUIVALENT CYLINDER

Two different forms of contact are identifiable.

(1) The contact such as between a ball and the race of a ball bearing, where the area of contact is nominally a point but, because of elastic deformation, becomes elliptical or, in the special case of a sphere and a flat plate, circular. This area of contact is sometimes referred to as the footprint.
(2) The contact, such as that between a cylindrical roller and a flat raceway, which produces an essentially rectangular footprint.

Most EHL rectangular contacts, such as that shown in Figure 7.1, can be represented with sufficient accuracy by an equivalent cylinder against a

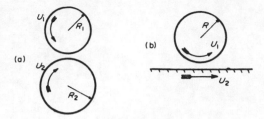

**Figure 7.1**  Equivalent cylinder

flat plane, where the radius of the equivalent cylinder is calculated as in
Hertzian contact (Section 2.2). This is achieved by putting

$$R = \frac{R_1 R_2}{R_1 + R_2} \tag{7.1}$$

where the surface velocities of the two components are preserved. In the case
of a concave surface the radius would be negative.

The same procedure is adopted for the elliptical footprint, where an
equivalent radius of curvature is calculated in the two principal directions,
$R_x$ being along the line of fluid entrainment and $R_y$ being perpendicular to it.


## 7.3  RIGID CYLINDER THEORY

The film shape between the cylinder and plane has the form

$$h = h_0 + R(1 - \cos \theta)$$

Solution of the Reynolds equation must clearly include the possibility of
cavitation in the divergent outlet region, as described in Section 6.13. An
approximation to the film shape is given by assuming a parabolic profile

$$h = h_0 + \frac{x^2}{2R}$$

Solutions were obtained for a parabolic cylinder (Martin, 1916; Purday,
1949) and for a circular cylinder (Floberg, 1961). Since Reynolds's equation
cannot really hold over the entire 90° inlet zone that Floberg considered,
the results of this analysis can only be regarded as accurate for relatively
thin films, where the pressure generated is largely concentrated in a small
area near the centre. In this case the solutions coincide. The load capacity
of the contact can be written

$$F_z = \beta(U_1 + U_2)\eta \frac{RB}{h_0}$$

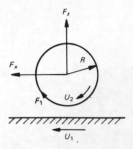

**Figure 7.2**   Cylinder and plane

The value of $\beta$ depends on $R/h_0$, but for $R/h_0$ greater than about $10^4$ may be taken as 2.45.

The sideways force on the cylinder due to the horizontal component of pressure is given by

$$F_x = 4.5\eta(U_1 + U_2)B\sqrt{\frac{R}{h_0}} \tag{7.2}$$

for large values of $R/h_0$. For the plane, of course, $F_x = 0$. If the real situation consisted originally of two discs, clearly there will be a horizontal force on each one which is given by

$$F_{x_1} = \frac{R}{R_1}F_x$$

$$F_{x_2} = \frac{R}{R_2}F_x$$

where $R_1$ and $R_2$ are the radii of the original cylinders.

The viscous drags along the plane surface and around the cylinder circumference are

$$F_1 = -\frac{F_x}{2} - \eta(U_1 - U_2)AB$$

$$F_2 = -\frac{F_x}{2} + (U_1 - U_2)AB \tag{7.3}$$

The value of $A$, assuming a full contribution from the cavitated region, is found to be 3.84 for cases where $R/h_0$ is large.

The forces described above are shown in Figure 7.2.

## 7.4   ELASTOHYDRODYNAMIC MECHANISM

When a load of significant size is imposed on a non-conforming contact, both surfaces will suffer elastic deflection to change the nominal point or line contact into a footprint having an area which is significant compared with the dimensions of the lubricant film. This produces a 'flat' on balls and rollers, with the load carried by a pressure spread over this flattened area in 'Hertzian' contact, as described in Chapter 2.

When the surfaces move, they will drag lubricant into the convergent inlet zone in the familiar hydrodynamic fashion. The lubricant will then pass through the flattened zone, to emerge in the divergent outlet, where cavitation will normally take place. For many cases the presence of the film does not significantly affect either the size or the shape of the flattened area and, indeed, the Hertzian pressure distribution can remain essentially unchanged.

As the lubricant passes through the gap, it is subject to an increasing pressure. Most fluids, with the notable exception of water, exhibit a marked increase in viscosity with increasing pressure. Indeed, in very-high-pressure contacts the viscosity can increase several hundredfold, rendering the fluid almost solid in its effect. The result of both these phenomena is drastically to increase the load-carrying abilities of the lubricated contact relative to that calculated by rigid body, isoviscous theory.

## 7.5  ELASTOHYDRODYNAMIC THEORY

To produce results of use to the designer of highly loaded contacts where viscosity increase and elastic deflection are important, the theory described above for rigid cylinders gives film thickness values which are severely underestimated. In practice the local elastic flattening of the curved surface spreads the load over a larger area and the viscosity increase enhances the film-building properties of the lubricant. To analyse this situation it is necessary to carry out the simultaneous solution of the Reynolds equation (7.4), the elastic deformation equation (7.5) and the equation relating viscosity to pressure (7.6):

$$\frac{\partial}{\partial x}\left(\frac{\rho h^3}{\eta}\cdot\frac{\partial p}{\partial x}\right) + \frac{\partial}{\partial y}\left(\frac{\rho h^3}{\eta}\cdot\frac{\partial p}{\partial y}\right) = 12U\frac{\partial(\rho h)}{\partial x} \tag{7.4}$$

$$v = \frac{1}{E'}\int_{x_1}^{x_2} p(s)\log(x-s)\cdot ds \tag{7.5}$$

$$\eta = \eta_0 e^{\alpha p} \tag{7.6}$$

If isothermal conditions no longer prevail, the energy equation and the conduction equation for the heat passing to the surfaces must be incorporated. The cavitation boundary condition must also be employed as described in sub-section 6.1.2.

Although a full solution had to wait until the introduction of the high-speed digital computer, inspired assumptions produced very good approximate solutions as early as 1949 in the work by Ertel (Grubin, 1949) and in the next decade by several workers (Petrusevich, 1951; Weber and Saalfeld, 1954). Even today analytical methods can give a valuable insight into the mechanisms at work (Archard and Baglin, 1986). However, the more accurate computer results, beginning with those of Dowson and Higginson in 1959 and extended and refined subsequently by many workers, are used as a basis for the design procedures described in this chapter.

## 7.6 RESULTS

### 7.6.1 Film Shape and Pressure Distribution

Results from the foregoing analyses permit the plotting of the shape of the distorted cylinder and the pressure distribution for a given set of conditions. Typically, for a relatively slow-speed, rectangular footprint these will have the form illustrated in Figure 7.3.

Considering first the film shape, we see the convergent inlet zone, followed by a virtually parallel section. At outlet there is a constriction which can amount to a reduction of 25% of the parallel film thickness.

The pressure curve is very close to the Hertzian dry contact pressure shown in Chapter 2. There is a build-up in the inlet and towards the outlet there is a pressure spike and the cavitation boundary. The spike is difficult to find experimentally, because it is very narrow (Kannel, 1965), but it can be shown theoretically to exist and is important because of the possibility of high subsurface stresses. There is evidence that, if a more accurate pressure–viscosity relationship is used (Roelands, 1966), the calculated severity of the spike may be reduced. Of course, the total area under the curve must be the same as that under the Hertzian half-ellipse. The length of the parallel section will be approximately $2\{(8RW/\pi BE')\}^{\frac{1}{2}}$ and the maximum pressure $(WE'/2\pi R)^{\frac{1}{2}}$.

As the speed increases, the pressure distribution will depart more and more from the Hertzian, as shown in Figure 7.4.

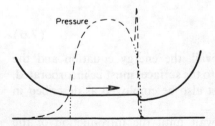

**Figure 7.3** Pressure distribution and film shape in EHL

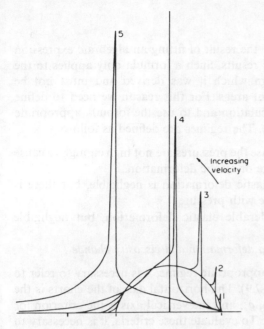

**Figure 7.4**  Effect of velocity on pressure distribution

For the elliptical footprint the results are very similar, as shown in Figure 7.5, which is an interference pattern for a circular EHL footprint. There is still an essentially parallel central section, while the constriction at outlet now extends around about 180° of the contact circle (Gohar and Cameron, 1966).

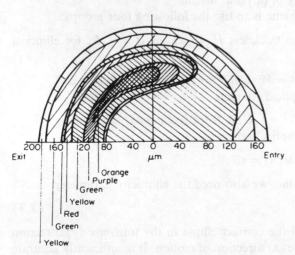

**Figure 7.5**  Interference pattern in elliptical contact, $k = 1$

### 7.6.2 Film Thickness

Formulae for film thickness are the result of fitting an algebraic expression to a large number of computed results. Such a formula only applies to the particular range of results from which it was derived and must not be extrapolated into wildly different areas. For this reason we need to define 'regimes' for film thickness calculation and to use the formula appropriate to each regime (Johnson, 1970). The regimes are defined as follows.

(1) *Rigid–isoviscous* In this case the pressures are not high enough to cause appreciable viscosity change or elastic deformation.
(2) *Rigid–piezoviscous* The elastic deformation is negligible, but there is significant viscosity increase with pressure.
(3) *Elastic–isoviscous* Considerable elastic deformation, but negligible viscosity change.
(4) *Full EHL solution, including deformation and viscosity change*

In order to determine the appropriate regime, it is necessary to refer to the relevant chart (Figures 7.6–7.9). The horizontal axis of the charts is the criterion for elastic deformation, $g_e$, and the vertical axis is the criterion for significant viscosity increase, $g_v$. To evaluate these criteria, it is necessary to examine the dimensionless groupings of the next sub-section.

### 7.6.3 Dimensionless Groups

The results from the computer analyses are conveniently presented in the form of dimensionless groups. This has the advantage of ease of plotting, independence of the system of units used and, if the groups are chosen intelligently, the possibility of physical insight.

The most popular scheme is to use the following four groups:

the dimensionless film thickness $\hat{H} = h_0/R$ or $\hat{H} = h_0/R_x$ for elliptical contacts

the load parameter $\hat{W} = W/E'RB$ for rectangular contacts or $\hat{W} = W/E'R_x^2$ for elliptical contacts

the speed parameter $\hat{U} = \eta_0 U/E'R$ for rectangular contacts or $\hat{U} = \eta_0 U/E'R_x$ for elliptical contacts

the materials parameter $G = \alpha E'$

For the elliptical contact we also need the ellipticity parameter

$$k = a/b \tag{7.7}$$

where $a$ is the semiaxis of the contact ellipse in the tranverse ($y$) direction and $b$ is the semiaxis in the ($x$) direction of motion. It is sufficiently accurate

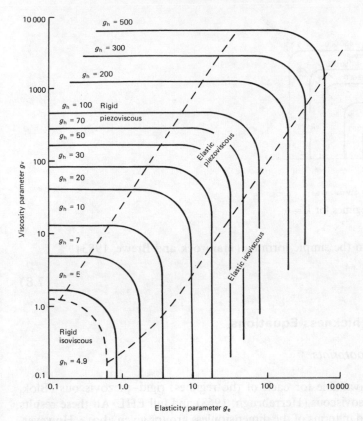

**Figure 7.6** Regimes of rectangular contact

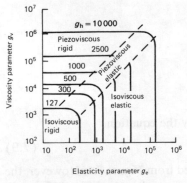

**Figure 7.7** Regimes for $k = 1$

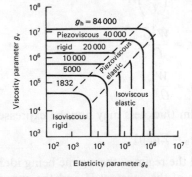

**Figure 7.8** Regimes for $k = 3$

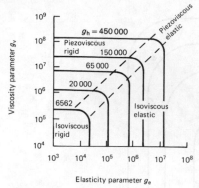

**Figure 7.9** Regimes for $k = 6$

to derive $k$ from the simple formula (Hamrock and Brewe, 1983)

$$k = \left(\frac{R_y}{R_x}\right)^{0.64}$$

(7.8)

### 7.6.4 Film Thickness Equations

*Rectangular Footprints*

Solutions are available for each of the regimes: rigid–piezoviscous (Blok, 1952), elastic–isoviscous (Herrebrugh, 1968) and full EHL. All these results can be expressed in terms of the dimensionless groups given above. However, it can be shown that only three groups are really necessary, thus rendering the results easy to plot. By choosing sensible groups, the relative effect of viscosity increase and elastic deflection may be seen.

If

$$g_h = \left(\frac{\hat{W}}{\hat{U}}\right)\hat{H}$$

$$g_v = \frac{\hat{W}^{\frac{3}{2}}\hat{G}}{\hat{U}^{\frac{1}{2}}}$$

$$g_e = \frac{\hat{W}}{\hat{U}^{\frac{1}{2}}}$$

the film thickness may now be expressed by the equation

$$g_h = Z g_v^m g_e^n$$

(7.9)

for all the regimes, the regime being identified from Figure 7.6. However, the values of the constant $Z$ and the exponents $m$ and $n$ are different for each regime, as indicated in Table 7.1.

**Table 7.1**

| Regime | Z | m | n |
|---|---|---|---|
| (1) | 2.45 | 0 | 0 |
| (2) | 1.05 | $\frac{2}{3}$ | 0 |
| (3) | 2.45 | 0 | 0.8 |
| (4) | 1.654 | 0.54 | 0.06 |

Note the very low value of the exponent $n$ in the full EHL solution. This indicates the effect of elastic modulus, since $g_e$ is the only term containing $E'$. This suggests that the film thickness is relatively insensitive to changes of elastic modulus and it is found to be relatively insensitive to changes of load. This is confirmed in practice, where an increase in load or decrease in modulus will produce a resulting increase in the flattening of the surfaces and a bigger effective load-carrying area. The above values are for the minimum film thickness at the constricted outlet. The film thickness over the essentially parallel region is given by the ratio

$$h_0/h_c = 0.72 - 0.81$$

The error incurred by assuming a ratio of 3/4 is negligible for most cases.

### Elliptical Footprints

These may be approached in the same way. The general equation is

$$g_h = Z g_v^m g_e^n f(k) \tag{7.10}$$

where

$$g_h = \hat{H}\left(\frac{\hat{W}}{\hat{U}}\right)^2$$

$$g_v = \frac{\hat{G}\hat{W}^3}{\hat{U}^2}$$

$$g_e = \frac{\hat{W}^{\frac{8}{3}}}{\hat{U}^2}$$

The 'constants' $Z$, $m$ and $n$ will be different for each regime, as will the function $f(k)$. The regime is identified from Figures 7.7–7.9. The values of $Z$, $m$ and $n$ are shown in Table 7.2.

Equation (7.10) gives the value of the minimum film thickness, i.e. at the outlet constriction. The ratio between this thickness and that in the parallel region shows considerably more variation than the elliptical case.

**Table 7.2**

| Regime | Z | m | n | $f(k)$ |
|--------|-----|-----|------|--------|
| (1) | 128 | 0 | 0 | $\gamma\phi^2\left[0.131\tan^{-1}\left(\dfrac{\gamma}{2}\right)+1.683\right]^2$ |
| (2) | 1.66 | $\frac{2}{3}$ | 0 | $[1-e^{-0.68k}]$ |
| (3) | 8.70 | 0 | 0.67 | $[1-0.85e^{-0.31k}]$ |
| (4) | 3.42 | 0.49 | 0.17 | $[1-e^{-0.68k}]$ |

## 7.7 DESIGN AIDS FOR EHL FILM THICKNESS

A valuable manual for the calculation of EHL film thickness in line contact is published as ESDU Item No. 85027 (1985). A similar publication is in preparation for three-dimensional (point) contacts.

### 7.7.1 Examples of Film Thickness Calculation

(1) A cylindrical roller in a roller bearing is 32 mm in diameter and 35 mm long. The outer race has a concave radius of 103.5 mm and it is this race that rotates at 100 rev/min. The load on the roller is 3 kN and it is lubricated with an oil of viscosity 0.003 Ns/m$^2$ and pressure viscosity coefficient $2.3 \times 10^{-7}$ m$^2$/N. Estimate the minimum film thickness in the contact.

First we need to calculate the equivalent cylinder radius, with $R_1 = 16$ mm and $R_2 = -103.5$ mm:

$$R = \frac{16(-103.5)}{16-103\cdot5} = 18.9 \text{ mm}$$

The linear speed of the two surfaces is the same, assuming no slip, and so

$$u = U = \frac{207 \times \pi \times 100 \times 2 \times \pi}{60} = 6.81 \text{ m/s}$$

Using the rigid, isoviscous solution and assuming that $R/h_0$ is going to be greater than $10^4$ gives

$$h = 2.45(U_1 + U_2)\eta RB/W$$

$$= 2.45 \times 2 \times 6.81 \times 3 \times 10^{-3} \times 18.9 \times 10^{-3} \times 35 \times 10^{-3}/3000$$

$$h = 0.022 \ \mu m$$

This is, of course, a very small film thickness, much smaller than the surface roughness of the roller, and this theory does not predict full film lubrication.

Now we shall introduce elastohydrodynamic theory into the problem.

Since the two surfaces are steel, $E' = E/(1 - v^2)$, and assuming $E = 208 \times 10^9 \, \text{N/m}^2$ and $v = 0.3$ gives $E' = 228 \times 10^9 \, \text{N/m}^2$. Calculating the operating parameters,

$$\hat{W} = 3000/(228 \times 10^9 \times 18.9 \times 10^{-3} \times 35 \times 10^{-3}) = 1.99 \times 10^{-5}$$

$$\hat{U} = 3000 \times 6.81/(228 \times 10^9 \times 18.9 \times 10^{-3}) = 4.74 \times 10^{-12}$$

$$\hat{G} = E' = 2.3 \times 10^{-7} \times 228 \times 10^9 = 524 \times 10^2$$

The criteria for the effect of viscosity change and elasticity are as follows

$$g_v = \frac{(1.99 \times 10^{-5})^{\frac{3}{2}} 524 \times 10^2}{(4.74 \times 10^{-12})^{\frac{1}{2}}} = 2137$$

$$g_e = 1.99 \times 10^{-5}/(4.74 \times 10^{-12})^{\frac{1}{2}} = 9.14$$

Plotting these values on Figure 7.6 shows a point just in the rigid–piezoviscous regime. Using the appropriate values from Table 7.1 in Equation (7.9),

$$g_h = 0.99(g_v)^{\frac{2}{3}} = 0.99(2137)^{\frac{2}{3}} = 164.2$$

$$\hat{H} = \frac{\hat{U}}{\hat{W}} g_h = \frac{4.74 \times 10^{-12}}{1.99 \times 10^{-5}} \times 164.2 = 391 \times 10^{-7}$$

$$h_0 = 391 \times 10^{-7} \times 18.9 \times 10^{-3} = 0.739 \, \mu\text{m}$$

Note that the film thickness has been increased by a factor of more than 30 by the inclusion of variable viscosity. This is a much more realistic thickness.

(2) A standard deep-groove ball bearing has balls of 8.5 mm diameter. The inner race rotates at 5000 rev/min. The groove in which the balls roll on the inner race has a radius in the direction of motion of 10 mm and a radius of curvature of 4.6 mm at right angles to the motion. The load on the bearing is 300 N and it is lubricated with an oil of viscosity $0.6 \, \text{Ns/m}^2$ and viscosity index $2.1 \times 10^{-8} \, \text{m}^2/\text{N}$. Estimate the film thickness.

The load on the most heavily loaded element in a rolling bearing is given as a pessimistic estimate by $5/N$ times the total load, where $N$ is the number of rollers or balls (Stribeck, 1901). If in this case there are 12 balls,

$$W = 5(300/12) = 125 \, \text{N}$$

The radii of curvature are

$$R = 4.25(10/(4.25 + 10)) = 2.98 \text{ mm}$$

$$R = 4.25(-4.6)/(4.25 - 4.6) = 55.8 \text{ mm}$$

$$k = (55.8/2.98)^{\frac{2}{3}} = 6.46$$

$$\hat{U} = \frac{0.6 \times 5.235}{228 \times 10^9 \times 2.98 \times 10^{-3}} = 4.623 \times 10^{-9}$$

$$\hat{W} = \frac{125}{228 \times 10^9 \times 2.98^2 \times 10^{-6}} = 6.174 \times 10^{-5}$$

$$\hat{G} = 2.1 \times 10^{-8} \times 228 \times 10^9 = 4.788 \times 10^3$$

This gives the criteria as

$$g_v = \frac{4.788 \times 10^3 \times 6.174^3 \times 10^{-15}}{4.623^2 \times 10^{-18}} = 5.27 \times 10^7$$

$$g_e = 2.786 \times 10^5$$

This is just in the piezoviscous, rigid regime on Figure 7.9, for $k = 6$. Using Equation (7.10) and the appropriate values from Table 7.2,

$$g_h = 1.66(5.27 \times 10^7)^{\frac{2}{3}}(1 - e^{-0.08 \times 6.46}) = 1.66 \times 1.406 \times 10^5(0.9876)$$

$$= 2.305 \times 10^5$$

$$\hat{H} = g_h \left(\frac{\hat{U}}{\hat{W}}\right)^2 = 2.305 \times 10^5 \left(\frac{4.629 \times 10^{-9}}{6.174 \times 10^{-5}}\right)^2 = 1.292 \times 10^{-3}$$

$$h_0 = 1.292 \times 10^{-3} \times 2.98 \times 10^{-3} = 3.85 \ \mu\text{m}$$

## 7.7.2  Inlet Shear Heating

As the fluid is dragged into the inlet zone, the shearing which takes place increases the temperature of the lubricant and, therefore, reduces its viscosity. This results in a reduction of film thickness, the magnitude of which is dependent on the parameter $I$, given by

$$I = \frac{\eta_0 U^2 \delta}{K}$$

where $\delta$ is the temperature–viscosity coefficient and $K$ is the thermal conductivity of the lubricant. For rectangular contacts and a fully flooded inlet, the film thickness as calculated from sub-section 7.5.4 must be multiplied by a thermal reduction factor obtained from Figure 7.10.

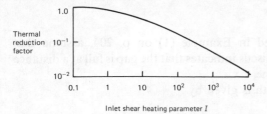

Thermal reduction factor

Inlet shear heating parameter $I$

**Figure 7.10**  Inlet shear heating effect

### 7.7.3  Starvation

All the results so far have assumed that the inlet to the contact is full of lubricant. If this is not the case, the contact is said to be starved. This results in a delay to the start of pressure build-up and a consequent reduction of film thickness.

For rectangular footprints the reduced film thickness, $h_s$, can be expressed approximately in terms of the fully flooded film thickness, $h_\infty$, by the formula

$$\frac{h_s}{h_\infty} = \frac{2}{\pi} \arctan[1.37(J + 0.5)^2]$$

where $J = b^{\frac{1}{3}} x_i/(2Rh_0)^{\frac{2}{3}}$ and $x_i$ is the distance from the edge of the Hertzian zone to the inlet boundary, which is normally considered to be the position where the films of lubricant on the two surfaces come together; and $b$ is the Hertzian half-width (Wymer and Cameron, 1974).

For the elliptical case we first determine the dimensionless distance $m_i$. As the inlet position is moved progressively inward from infinity, it reaches a position where it begins to affect the film thickness. This distance from the centre, made dimensionless by dividing by $b$, the Hertzian half-width, is $m_i$. If $h_\infty$ is the minimum film thickness in the fully flooded case,

$$m_i = 1 + 3.34 \left[ \frac{R_x^2 h_\infty}{b^2} \right]^{0.56}$$

and we can then calculate the dimensionless minimum film thickness for the starved condition from

$$\frac{h_s}{h_\infty} = \left( \frac{m - 1}{m_i - 1} \right)^{0.25}$$

where $m$ is the actual dimensionless distance to the inlet boundary (Hamrock and Dowson, 1981).

Clearly, in a case of a starved contact the effect of the shear heating in the inlet is reduced. The two effects are, therefore, interdependent. A useful review of the combined effect can be found in Gohar (1988).

### Examples of Starved Contacts

(1) Taking the roller described in Example (1) on p. 204, assume that observation of the inlet meniscus indicates that the gap is full at a distance of 5 mm from the centre line.

The Hertzian half-width is given by

$$b = \left[ \frac{8WR}{\pi BE'} \right]^{\frac{1}{2}}$$

$$= \left[ \frac{8 \times 3000 \times 18.9 \times 10^{-3}}{\pi \times 35 \times 10^{-3} \times 228 \times 10^3} \right]^{\frac{1}{2}}$$

$$b = 1.35 \times 10^{-4} \text{ m} = 0.135 \text{ mm}$$

Therefore, the distance of the inlet boundary from the edge of the Hertzian zone is

$$x_i = 5 - 0.135 = 4.865 \text{ mm}$$

$$J = \frac{0.135^{\frac{1}{3}} \times 4.865}{(2 \times 18.9 \times 1.23 \times 10^{-3})^{\frac{2}{3}}} = 19.3$$

$$\frac{h_s}{h_\infty} = \frac{2}{\pi} \arctan(1.37 \times 19.8^2) = 0.9988$$

In other words, the film thickness is virtually unaffected by the delay in pressure build-up.

Only if the contact suffers much more starvation than this will the film thickness be diminished. For example, if the inlet meniscus were 0.5 mm from the centre line,

$$x_i = 0.5 - 0.135 = 0.365 \text{ mm}$$

$$J = \frac{0.135^{\frac{1}{3}} \times 0.365}{(2 \times 18.9 \times 1.23 \times 10^{-3})^{\frac{2}{3}}} = 1.448$$

$$\frac{h_s}{h_\infty} = \frac{2}{\pi} \arctan(1.37 \times 1.948^2) = 0.88$$

Even in this case only 12% of the film is lost.

(2) Using Example (2) on p. 205, let us suppose that there is a film adhering to each surface of thickness 1 $\mu$m as they enter the contact. If we neglect the film thickness and the flattening at the contact, simple geometry shows that the film will be complete at a distance $x$ from the centre line, given by $x = (2Rh)^{\frac{1}{2}}$, where $h$ is the film thickness at the inlet meniscus – in this case $2 \times 10^{-3}$ mm. This is, of course, only a guess at where the pressure build-up begins.

$$x = (2 \times 2.98 \times 2 \times 10^{-3})^{\frac{1}{2}} = 0.109 \text{ mm}$$

For an elliptical contact, the Hertzian half-width, $b$, is given with sufficient accuracy by

$$b = \left[ \frac{6ZWR_x}{\pi k E'} \right]$$

where

$$Z = 1.0003 + \frac{0.5968}{R_y/R_x}$$

$$= 1.0003 + \frac{0.5968}{55.8/2.98} = 1.032$$

$$b = \left( \frac{6 \times 1.032 \times 125 \times 2.98}{\pi \times 6.46 \times 228 \times 10^3} \right)^{\frac{1}{3}} = 0.0793$$

The maximum distance for starvation to have a noticeable effect is, therefore,

$$m_i = 1 + 3.34 \left( \frac{2.98 \times 1.94 \times 10^{-5}}{0.0793^2} \right)^{0.56} = 1.242$$

The actual dimensionless distance is

$$m = 0.109/0.0793 = 1.37$$

Since $m > m_i$, no starvation effect will be seen.

If the film were reduced to 0.75 μm on each surface, the inlet meniscus would move to

$$x = (2 \times 2.98 \times 1.5 \times 10^{-3})^{\frac{1}{2}} = 0.0945$$

giving

$$m = 0.0945/0.0793 = 1.19$$

In this case $m < m_i$, and so the reduced film thickness is

$$\frac{h_s}{h_\infty} = \left( \frac{1.19 - 1}{1.242 - 1} \right)^{0.25} = 0.94$$

a reduction of 6%.

### 7.7.4 Effect of Film Thickness on Fatigue Life

In many rolling contacts, especially rolling element bearings, the primary cause of failure is fatigue due to the repeated application and removal of stress on the material as it passes through the contact. It is not immediately obvious how the presence of a lubricant film helps to delay the fatigue failure, as these stress cycles are still present. However, the improved life achieved with a 'thick' lubricant film can be explained in several ways.

The Hertzian pressure profile is smoothed by the lubricant, so that the application of the stress is not as sudden: see Figure 7.4. The spike can be viewed as a cause of stress, but it seems unlikely that it is nearly as pronounced in practice as in these theoretical profiles.

In all rolling contacts microslip takes place (Halling, 1976), resulting in shear stresses at the surface. These combine with the contact stresses already present, to move the maximum shear stress, previously below the surface, closer to the surface, which is where cracks can be initiated. The lubricating effect of the fluid will greatly reduce these stresses.

Fatigue cracks will often propagate from surface defects. A thick film will prevent asperity contact and its associated microscopic stress fields close to these defects.

The effectiveness of the film is quantified by the specific film thickness, $\lambda$, where $\lambda = h/R_a$ and the mean roughness, $R_a$, is the square root of the sum of the squares of the two surface roughnesses. Harris (1984) gives a value of $\lambda \approx 3$ for a complete film for rolling bearings. As $\lambda$ is reduced below this value, the fatigue life will be progressively reduced.

## 7.8 TRACTION

In EHL, as in all lubrication mechanisms, including dry rolling, surface tractions are present. Even if pure rolling is taking place, energy is required to compress the fluid as it enters the contact, the only possible source of this energy being the motion of the surfaces. This will be seen as a retarding force on the surfaces, which we will call the rolling friction, $F_R$.

If the surfaces are moving at different speeds, the slower surface will attempt to retard the faster one and vice versa, this constituting the sliding friction, $F_S$. Hence, the retarding force on each surface may be written

for the slower surface, $F_R - F_S$

for the faster surface, $F_R + F_S$

Typical traction curves are shown in Figure 7.11, from which several points can be noted.

(1) At constant rolling speed $F_R$ is virtually unaffected by load, but rises with increasing rolling speed.
(2) For low sliding speeds $F_S$ is proportional to sliding speed, indicating an essentially Newtonian behaviour.
(3) At constant rolling speed $F_S$ increases with load. This is to a small extent due to the reduction of film thickness, but to a much greater extent to the increase of viscosity with pressure.
(4) At constant load $F_S$ decreases with increasing rolling speed. This is because of the increase of film thickness.

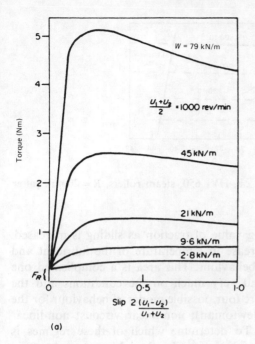

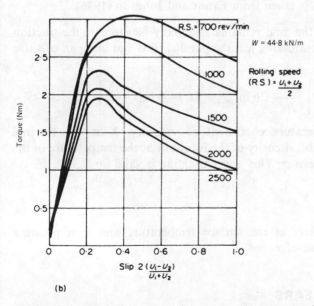

**Figure 7.11**   Traction curves for rectangular contact: four-disc machine results (Dowson and Whomes, 1967)

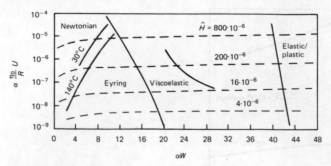

**Figure 7.12** Regimes of traction: oil, HVI 650; steam rollers, $R = 20$ mm (after Evans and Johnson, 1986)

(5) There seems to be a limiting value of traction as sliding is increased. This limit is due to the increase in temperature of the lubricant and departure from Newtonian behaviour. This area is a complicated one and performance will depend very much on the conditions and the individual lubricant. There are four possible modes of behaviour for the lubricant: linear viscous (Newtonian); non-linear viscous; non-linear viscoelastic; elastic–plastic. To determine which of these regimes is appropriate, it is necessary to plot a map for the lubricant as shown in Figure 7.12, which is taken from Evans and Johnson (1986).

For a case where the fluid remains essentially Newtonian, the traction can be calculated by considering just the parallel zone and using an effective viscosity, given by

$$\bar{\eta} = \frac{4K}{b\delta(U_2 - U_1)^2}\left[\frac{\alpha W}{B} + 2b\ln(U_2 - U_1) + b\ln\left(\frac{\eta_s\delta}{2K}\right)\right]$$

where $\delta$ is the temperature coefficient of viscosity, $K$ is the thermal conductivity and $\eta_s$ is the viscosity of the lubricant at the temperature of the surfaces at ambient pressure. This approximation is valid for

$$\eta_x\delta\frac{(U_2 - U_1)^2}{\delta K} \gg 1$$

where $\eta_x$ is the viscosity at the surface temperature and at a pressure appropriate to the value of $x$.

## 7.9 CAMS AND GEARS

The discussion so far has been concerned with the sort of contact encountered in rolling element bearings, wheels, etc. The situation with cams and gears can be somewhat different, with the inclusion of large velocities of approach and other dynamic effects. Some of the film thickness formulae suggested for these cases are given below.

For involute gears the contact at the pitch line may be modelled as two rollers, having radii of curvature $D_1 \sin \phi/2$ and $D_2 \sin \phi/2$, with the rollers rotating at the same angular velocities as the gearwheels themselves.

Holmberg (1982) gives the pitch-line film thickness, $h_p$ in microns, for gears on parallel shafts as

$$h_p = 1.7 \times 10^{-3}(\eta_0 V_p)^{\frac{2}{3}} l^{\frac{1}{3}}$$

where $V_p$ is the velocity at the pitch line in m/s, $\eta_0$ is the viscosity in cP and $l$ is the distance between centres in mm. In this case a standard value has been taken of the pressure–viscosity coefficient and the piezoviscous regime has been assumed.

An earlier approximation is given in *The Tribology Handbook* as

$$h_p = 3.5 \sqrt{V_e R \eta_0}$$

where $V_e$ is the entraining velocity in m/s, $R$ is the relative radius of curvature on the pitch line in metres and $\eta_0$ is the viscosity in poise.

$$R = \frac{D_1 D_2 \sin \phi}{2(D_1 + D_2) \cos^2 \phi}$$

and

$$V_e = 0.4V \sin \phi \cos \sigma$$

where $\phi$ is the normal pressure angle and $\sigma$ is the helix angle.

The minimum film thickness encountered during the contact is $h_p$ for gears of approximately equal size, but reduces to around 50% of $h_p$ for large gear ratios.

Experience suggests that, in order completely to avoid surface distress, we need a specific film thickness, $\lambda$, of more than 2. However, the situation is so complicated by running-in and transient effects that results appear like those shown in Figure 7.13, from Tallian (1967).

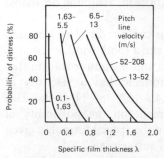

**Figure 7.13** Probability of surface distress

## 7.10 PROBLEMS

1 Calculate the minimum film thickness for a steel ball of 30 mm diameter rolling along a flat steel plate at a linear speed of 3 m/s. It can be regarded as fully flooded with oil of viscosity 80 cP and pressure viscosity coefficient $3 \times 10^{-8}$ m$^2$/N. It is loaded only by its own weight. (Steel weighs 78 kN/m$^2$.)

2 Repeat Problem 1 with the lubricant changed to water. Water has a viscosity of 0.6 cP and the pressure viscosity coefficient may be taken as zero.

3 A cylindrical rubber tyre of 800 mm diameter and 250 mm wide rolls along a flat road at 50 km/h under a vertical load of 2 kN. Assuming that a film of water is present on the road to a depth of 2 mm, estimate the film thickness between tyre and road. For rubber $E = 4$ N/mm$^2$ and Poisson's ratio $= 0.5$.

4 A steel strip 60 cm wide is transported on cylindrical rollers of 10 cm diameter rotating at 600 rev/min. At any instant the strip is carried by ten rollers, each carrying 100 N. If the contacts are fully lubricated by an oil of viscosity $5 \times 10^{-2}$ Ns/m$^2$ and the strip moves at 50 cm/s, calculate the minimum film thickness at the contacts, the traction force on the strip and the energy consumption of the rollers.

5 Pushing the theory to its limit, calculate the film thickness in a journal bearing, 50 mm long, having a shaft diameter of 65 mm and a diametral clearance of 1 mm, lubricated with an oil of viscosity 90 cP and coefficient $8 \times 10^{-8}$ m$^2$/N. The journal is steel and the bush is of plastic having $E = 200$ N/mm$^2$ and Poisson's ratio $= 0.33$. The shaft rotates at 600 rev/min. Are there any reasons why the theory is not applicable to this case?

6 A cylindrical roller of 30 mm diameter and 30 mm long rides with its axis horizontal on a well-lubricated metal foil, which is inclined at an angle of 1° to the horizontal. The foil is stretched between rollers which drive it at a speed of 120 mm/min in the upward direction. The roller rotates about its axis without slip relative to the foil, while otherwise remaining fixed in space. Estimate the viscosity of the lubricant, assuming negligible elastic deflection of the foil.

7 Two meshing involute spur gears have pitch circle radii of 60 mm and 120 mm, respectively, a pressure angle of 20° and a tooth width of 20 mm. The speed of the larger wheel is 1500 rev/min and the power transmitted is 40 kW. The gears are steel and the lubricant has a viscosity of 0.08 Ns/m$^2$ and a pressure coefficient of $2.4 \times 10^{-8}$ m$^2$/N. If the c.l.a. roughness of each surface is 0.4 $\mu$m, comment on the possibility of satisfactory operation of the pair. Use each of the special formulae given in Section 7.9 and also the general theory of sub-section 7.5.4.

8 A steel ball of 20 mm diameter rolls at a speed of 2 m/s in a straight steel

groove of 12 mm radius, carrying a load of 30 N. What is the minimum thickness of the lubricant film on each surface, to avoid starvation effects? If the inlet were fully flooded, would inlet shear heating be important? The lubricant has a viscosity of 95 cP, a pressure–viscosity coefficient of $3 \times 10^{-8} \, \mathrm{m^2/N}$ and a temperature–viscosity coefficient of $0.575/°C$, and the thermal conductivity of the fluid is $0.13 \, \mathrm{Wm/m^2 \, °C}$. Note that although Figure 7.10 is derived for rectangular contacts, an indication of the effect of inlet shear heating in elliptical contacts may be derived from it.

9  In Section 7.8 two little formulae are given for the tractions on the two surfaces. Explain why these do not contravene Newton's Third Law and draw sketches showing all the forces acting on the two rollers.

# Chapter 8
# Bearing Selection

## 8.1 INTRODUCTION

In the preceding chapters we have examined many kinds of tribological contact. Advice has been given on the theory and design of the types of bearing available, but we have yet to establish the criteria used to decide which solution to adopt for any tribological problem. In order to select the appropriate bearing, we must first consider the options available.

## 8.2 BEARING TYPES

The possible main bearing types are listed below.

(1) Plain rubbing bearing with a sensible choice of compatible materials.
(2) As (1), but including a solid film either attached to a surface or interposed.
(3) Rolling elements inserted between the surfaces.
(4) A fluid film maintained between the surfaces by an external pressure source (hydrostatic lubrication).
(5) A fluid film maintained between the surfaces by the motion of the contact (hydrodynamic lubrication).
(6) A magnetic field to separate the surfaces.
(7) The use of a mechanism or flexible material in bearings of limited travel. In this case the tribological problem has been designed out.

## 8.3 LIMITS OF TRIBOLOGICAL SOLUTIONS

The nature of the limitations encountered is shown in Figure 8.1

*Strength considerations* Considering the geometry of the contact and the materials used, we can define the maximum load which may be safely

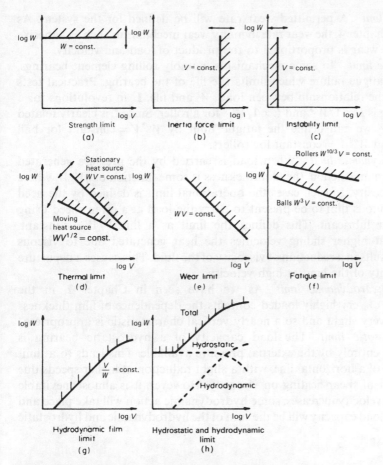

**Figure 8.1** Load and speed limits on performance

applied to the contact. Thus, for a cylindrical or spherical contact Hertzian theory will define the safe working load, while for other geometries even simpler calculations will be appropriate.

*Inertia limit* Since tribological components are in motion, they are subject to stresses arising from their inertia. This defines a maximum speed above which failure will ensue. This could be the bursting of a shaft due to centrifugal force at high rotational speed or the combined centrifugal and gyroscopic forces in rolling element bearings, the bursting of the cage or similar effects.

*Frictional instabilities* At low operating speeds some tribological systems suffer from stick–slip instability.

*Thermal effects* The heat produced by friction and fluid shear will cause a temperature rise in the components. There will be a maximum permissible temperature defined somewhere in the system.

*Wear limit*   A permitted wear rate will be defined for the system. As shown in Chapter 4, the wear rate in many wear mechanisms such as adhesive or abrasive wear is proportional to the product of load and velocity.

*Fatigue limit*   In some mechanisms, notably rolling element bearings, it is often fatigue failure which limits the life of the bearing. Practical tests show that the relationship between load, $W$, and life, $L$, in revolutions for a ball bearing is $L \propto 1/W^3$ and $L \propto 1/W^{\frac{10}{3}}$ for a roller. Since $L$ is clearly related to speed, $V$, we can define the fatigue limit by $W^3 V = $ constant for ball bearings and $W^{\frac{10}{3}} V = $ constant for rollers.

*Hydrodynamic limit*   When load is carried by the pressure generated by hydrodynamic action, the film thickness is some function of $\eta V/W$, where $\eta$ is the viscosity. In such cases the operational limit is defined by the need for a continuous film to be present to carry the load at a given speed using a particular lubricant. This defines the limit as a line $V/W = $ constant. However, at higher sliding velocities the heat generated due to viscous shearing results in a reduction of viscosity of the fluid. This causes a departure from linearity of the limit at high velocity.

*Elastohydrodynamic limit*   As we have seen in Chapter 7, in the lubrication of very highly loaded contacts the dependence of film thickness on load is very slight and so a nearly vertical characteristic is appropriate.

*Hydrostatic limit*   The load capacity of a hydrostatic bearing is determined entirely by the external pressure available. This leads to a limit in the form of a horizontal line, with a slight reduction at higher speeds due to the effect of shear heating on viscosity. However, it is almost inevitable that, as the velocity increases, some hydrodynamic action will take place and so the total load capacity will be the sum of the hydrodynamic and hydrostatic effects.

## 8.4   SELECTION OF JOURNAL BEARINGS

In order to see how the above limits may be combined to produce the characteristic limit curve for a bearing solution, we consider three journal bearing types: the dry rubbing bearing based on PTFE compounds, the rolling contact bearing and the hydrodynamic journal bearing. These characteristics are shown in Figure 8.2.

If we plot these characteristics for different shaft sizes we obtain a chart as shown in Figure 8.3. These curves cover the majority of the range of engineering applications. From the plot we can see clearly the advantages of rolling bearings at speeds in the range 1000–2000 rev/min, which explains their use in such situations – e.g. small electric motors. With larger shafts the enhanced load capacity of hydrodynamic bearings is clearly demonstrated. This explains why these bearings are so often used in such applications as

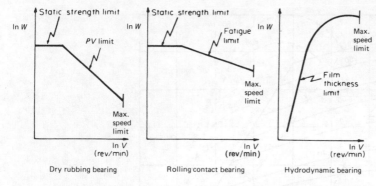

**Figure 8.2** Limits on three types of journal bearing

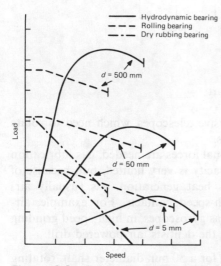

**Figure 8.3** Comparison of three bearing types

large steam turbines, where it is common practice to use a form of hydrostatic jacking to preclude metallic contact during start-up and stopping.

A more complete chart for journal bearing selection based on a nominal life of 10 000 h, assuming a viscosity grade of 32–100 (BS 4231) and a length/diameter ratio of unity, except for the rolling bearings, is shown in Figure 8.4. This is based on Neale (1967) which forms the basis of ESDU Item No. 65007 (1965). Externally pressurised bearings will operate over the entire range up to shaft-bursting speed. These give very high stiffness and low friction torque, especially at start-up and low speeds, and have a load capacity only limited by the size of the bearing and the pump pressure. However, because of the complexity of design and cost of installation, they are only used where these properties are essential. For example, externally

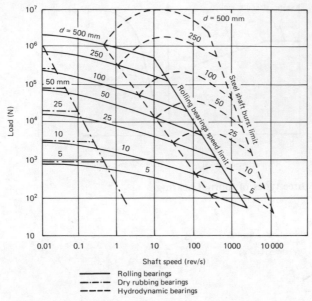

**Figure 8.4** Journal bearing selection chart

pressurised bearings are found on massive telescopes, which need to rotate with great precision at very low speeds.

For cases where even lower frictional forces are needed, gas lubrication can be used. In this case the load capacity is very limited, but, because of the low friction and consequent low heat generation, gas (usually air) lubrication can be used for very-high-speed contacts. For example, air-bearings are used in instruments, such as gyroscopes, in high-speed grinding and drilling spindles and, of course, in the dentists' air-powered drill.

*Example* Find a suitable bearing for a 50 mm diameter shaft, rotating at 1500 rev/min and carrying a load of 10 kN.

1500 rev/min = 25 rev/s. Enter Figure 8.4. The point indicated is well above the 50 mm line for rubbing bearings, and so these are not viable. It lies just above the rolling bearing line. A rolling bearing is possible, but nothing is in reserve on fatigue life. The point lies well below the line for hydrodynamic bearings, indicating that such a bearing would be very suitable. The final bearing selection will also involve practical considerations, such as space available, lubricant supply, temperature and environmental effects, etc.

## 8.5 SELECTION OF THRUST BEARINGS

Similar considerations may be applied to the selection of thrust bearings. The results are summarised in Figure 8.5, which is based on Neale (1967)

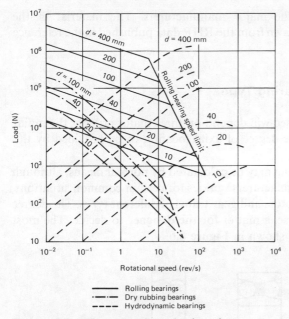

**Figure 8.5**   Thrust bearing selection chart

which reappears in ESDU Item No. 67033 (1967). This chart is based on a nominal life of 10 000 h and, with the exception of the rolling element bearings, assumes a 'typical' ratio between outside and inside diameter. This ratio will depend on the particular application, but is unlikely to exceed 2. Theoretically, higher load capacity can be obtained by increasing the outer diameter, but tolerance to misalignment would be less and excessive heating could impair the performance.

*Example*   Select a thrust bearing type to carry a load of 10 kN at 1500 rev/min if the shaft size is 50 mm.

1500 rev/min = 25 rev/s. Entering Figure 8.5 with 10 kN and 25 rev/s gives a point which is almost on the speed limit for rolling element bearings, but is well under the interpolated line for hydrodynamic bearings.

## 8.6   SELECTION OF ROLLING ELEMENT BEARINGS

It is most unlikely that the reader will ever be called on to design a rolling element bearing, but the selection of such a bearing is a common design task. Each bearing manufacturer produces a guide to the selection and use of his particular range of bearings. For more than a decade all the major manufacturers world-wide have conformed to the recommendations of ISO 281/1:1977 (BS 5512 Pt 1:1977). This has been reviewed recently to reflect the widespread availability of cleaner steels, but the selection procedures remain very similar for the major manufacturers. The material in the discussion that follows is drawn from the RHP data published in the reference given (RHP, 1977).

### 8.6.1   Selection of Bearing Type

The type of bearing to be used will depend on three major criteria: the radial load, the axial load and the degree of axial location to be provided by the bearing.

The loads on the bearing may be calculated by normal means, although help is provided in the manufacturers' guides for certain common situations, such as shafts with heavy rotors and gear trains of different types. The degree of axial location is, of course, a matter for the designer to decide. The most common bearing types are shown in Figure 8.6.

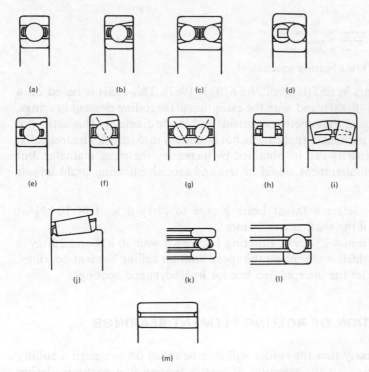

**Figure 8.6**   Rolling bearing types

*(a)   Single-row Radial Ball Bearings – Deep Groove*

*Radial load*   Good capacity.
*Axial load*   Moderate capacity.

*Axial location* Both directions, with a little movement.
*Comments* Versatile, with a wide range of load, speed and mounting conditions.

### (b) Single-row Radial Ball Bearing – Filling-slot Type

*Radial load* Greater than the deep-groove equivalent.
*Axial load* Not for use with axial load only. Axial load should be less than 20% of radial in order to prevent damage to the balls by the edges of the filling slots.
*Axial location* Both directions, with a little movement.
*Comments* In this bearing slots are cut in the raceways to allow more balls to be inserted into the bearing.

### (c) Double-row Radial Ball Bearings

*Radial load* Higher than the single-row bearing.
*Axial load* Not for axial load only. Axial load limited to 20% of radial load.
*Axial location* Both directions, with a little movement.
*Comments* More susceptible to misalignment than single-row.

### (d) Double-row Self-aligning Ball Bearings

*Radial load* Moderate.
*Axial load* Low capacity, owing to the wedge action of balls in spherical outer race.
*Axial location* Both directions, with some movement.
*Comments* Used where there is a need for misalignment allowance.

### (e) Single-row Separable Ball Bearings – Magneto Bearings

*Radial load* Low.
*Axial load* Low.
*Axial location* One direction only.
*Comments* The bearing races are detachable and interchangeable. They are often used in pairs, one bearing being adjusted against the other.

### (f) Single-row Angular Contact Ball Bearings

*Radial load* Good.
*Axial load* Good.
*Axial location* One direction only.
*Comments* Better axial load capacity than deep-groove balls. Often used in pairs, face-to-face or back-to-back.

### (g)  Double-row Angular Contact Ball Bearings

*Radial load*   Good.
*Axial load*   Good in both directions, except in the type with filling slots.
*Axial location*   Both directions, with some movement.

### (h)  Single-row Cylindrical Roller Bearings

*Radial load*   Better than the equivalent ball bearing.
*Axial load*   None for basic types. Some bearings include specially designed ribs to carry very small, intermittent axial force. These rely on effective lubrication and mounting design.
*Axial location*   None for basic types.

### (i)  Double-row Spherical Rollers

*Radial load*   Very good.
*Axial load*   Not for use with axial load only. Axial load is limited to 50% of the radial load.
*Axial location*   Both directions, with some movement.
*Comments*   For use where allowance for misalignment is desirable.

### (j)  Taper Roller Bearings

*Radial load*   Good.
*Axial load*   Good in one direction.
*Axial location*   In one direction.

### (k)  Single-row Thrust Ball Bearings – Flat Race

*Radial load*   None.
*Axial load*   Light loads only.
*Axial location*   One direction only.
*Comments*   The flat races give low friction, but correspondingly low load capacity.

### (l)  Single-row Thrust Ball Bearings – Grooved Race

*Radial load*   None.
*Axial load*   Good.
*Axial location*   One direction only.
*Comments*   The effect of centrifugal force limits the speed range.

*(m) Needle Rollers – Thrust and Journal*

*Comments* The rollers have a high length/diameter ratio. They are often used in situations where there is not sufficient space for the equivalent ball or roller bearing.

## 8.6.2 Selection of Bearing Size

Having decided on the bearing type, the size must be determined. For each bearing the manufacturer will quote:
C – the dynamic load capacity of the bearing
$C_o$ – the static load capacity of the bearing
Each of these must be checked against the operating conditions.

*Dynamic Capacity*

The dynamic failure of rolling bearings takes place through various fatigue mechanisms. It is in the nature of fatigue that the life of samples shows a large scatter and can only be expressed statistically. The 'life' of the bearing is usually defined by the $L_{10}$ life, which is the number of millions of revolutions that 90% of the bearings are expected to exceed before failure. If $P$ is the dynamic equivalent load, which is in most cases the applied load while running (but check with the manufacturer's recommendations), the life is given by

$$L_{10} = \left(\frac{C}{P}\right)^3 \text{ for ball bearings}$$

$$L_{10} = \left(\frac{C}{P}\right)^{\frac{10}{3}} \text{ for roller bearings}$$

Or, for a constant speed of $n$ rev/min,

$$L_{10} = \frac{16\,667}{n}\left(\frac{C}{P}\right)^3 \text{ hours for ball bearings}$$

$$L_{10} = \frac{16\,667}{n}\left(\frac{C}{P}\right)^{\frac{10}{3}} \text{ hours for roller bearings}$$

Modifications to this life should be made to get a more realistic life, $L$, by

$$L = a_1 a_2 a_3 L_{10}$$

*Reliability factor $a_1$* There may be some applications where 90% reliability is not good enough. In this case we introduce the factor $a_1$, as given in the table below.

| Reliability (%) | Life factor $a_1$ |
| --- | --- |
| 90 | 1.0 |
| 95 | 0.62 |
| 96 | 0.53 |
| 97 | 0.44 |
| 98 | 0.33 |
| 99 | 0.21 |

*Material and lubrication factor $a_{2,3}$*   This factor is used as the product of $a_2$ and $a_3$. It is used to include the improved materials ($a_2$) and the lubricating and operating conditions ($a_3$). The procedure for calculating $a_{2,3}$ is given as follows:

(1)  Calculate the basic $L_{10}$ life.
(2)  Obtain the recommended lubricant viscosity at 37.8 °C from Figure 8.7.

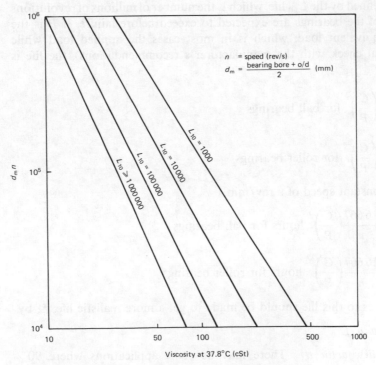

**Figure 8.7**  Choice of lubricating oil for rolling bearing

(3) Determine the actual lubricant viscosity at 37.8 °C.
(4) Divide the viscosity in (3) by that in (2), to give the viscosity ratio, $V_r$.
(5) Use Figure 8.8 to obtain the viscosity ratio at the operating temperature, $V_t$. For speeds given by $(d_m n)$ less than 10 or $V_r$ less than 1, use $V_t = V_r$.
(6) Read off the factor $a_{2,3}$ from Figure 8.9.

This procedure is valid for normal temperatures and lubricants and assumes a viscosity index greater than 90. Above 150 °C there is a loss of bearing hardness, which will adversely affect bearing life, and for non-conventional lubricants, such as water–oil emulsions, further reductions in life are to be expected.

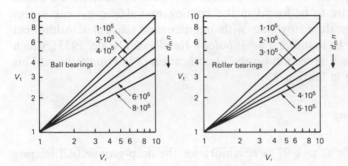

**Figure 8.8** Viscosity ratio at operating temperature

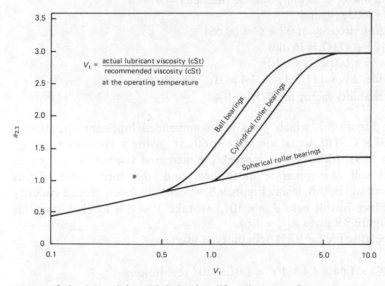

**Figure 8.9** Material and lubrication life adjustment factor, $a_{2,3}$

## Static Loading

The manufacturers' tables give a static load capacity $C_0$. This is the load which, when applied to the stationary bearing, produces a permanent deflection of 0.0001 of the rolling element diameter. This is known as brinelling. For satisfactory performance the maximum equivalent static load, which is in most cases the actual value of the static load, must not exceed $C_0$.

## Other Considerations

The preceding data assume that the bearing is installed correctly, with the appropriate tolerances, and is adequately lubricated. Instructions on fitting and lubrication are to be found in the bearing manufacturers' catalogues and technical literature, together with maintenance schedules. Additional information is to be found in *The Tribology Handbook* (Neale, 1973), which includes an interesting section on the identification of bearing failures. This is also dealt with in Section 9.9.

### 8.6.3 Example

Determine the life to give 97% reliability for the deep-groove ball bearing described below.

> $C = 11.2$ kN; load, $P = 1$ kN
> bore diameter $= 25$ mm; outside diameter $= 47$ mm
> speed $= 600$ rev/min
> lubricant viscosity at $37.8°C = 60$ cSt
> $d_m = (25 + 47)/2 = 36$ mm
> $d_m n = 36 \times 600 = 216 \times 10^2$
> basic life, $L_{10} = (11.2/1)^3 = 1.4 \times 10^9$ revolutions
> The reliability factor for 97% = 0.44.

Enter Figure 8.7, which gives a recommended lubricant viscosity of 55 cSt at 37.8°C. The actual viscosity is 60 cSt, giving a viscosity ratio, $V_r$, of $60/55 = 1.091$. Although, in general, the increased viscosity will enhance the life, it will also generate more heat and, therefore, some of this enhancement will be lost. Using Figure 9.8, we can calculate a revised viscosity ratio, $V_t$. Since in this case $d_m n < 10^5$, we take $V_t = V_r = 1.091$. Using this value in Figure 8.9 gives $a_{2,3} = 1.64$.

The bearing life for 97% reliability is given by

$$L = 0.44 \times 1.64 \times 1.4 \times 10^9 = 1.01 \times 10^9 \text{ revolutions}$$

## 8.7 PROBLEMS

1  Select a suitable type of journal bearing to carry a radial load of 100 kN at 6000 rev/min, assuming a shaft diameter of 100 mm.

2  A journal bearing of shaft diameter 25 mm is to carry a load of 9 kN at a speed of 30 rev/min. Decide on a suitable bearing type.

3  Select a thrust bearing type to carry 300 N at 6000 rev/min, if the shaft diameter is 10 mm.

4  A thrust bearing of 400 mm shaft diameter is to carry a load of 500 N at 30 rev/min. Choose a suitable bearing type.

5  A single-row cylindrical roller bearing having a dynamic capacity of 523 kN, a bore of 110 mm and an outside diameter of 280 mm is lubricated by an oil having a viscosity of 40 cSt at 37.8°C. Calculate the life for a reliability of 96% if the speed of rotation is 1500 rev/min and the load is 20 kN.

6  Determine the necessary value of $C$ for a ball bearing carrying a radial load of 20 kN on a shaft of diameter 50 mm rotating at 1000 rev/min, lubricated with an oil of viscosity 50 cSt at 37.8°C, to give an $L_{10}$ life of 1000 h.

7  A single-row roller bearing having a bore of 160 mm, an outside diameter of 340 mm and a value of $C$ of 1.32 MN has been selected to carry a load of 200 kN at 260 rev/min. The lubricant has a viscosity of 40 cSt at 37.8°C. The required life is $65 \cdot 10^{7}$ revolutions. Comment on the choice of bearing and lubricant.

# Chapter 9
# Lubricating Systems

## 9.1  INTRODUCTION

In the preceding chapters we have seen the beneficial effect of the presence of the lubricant on the bearing performance. The lubricant has three main functions: (1) to provide a coherent film between the surfaces; (2) to remove heat from the contact; (3) to prevent the ingress of dirt and other contaminants. The following sections deal with the volume and means of lubricant supply, the monitoring of the system and the lubricant, and the likely results of poor performance in this area.

## 9.2  FLOW REQUIREMENT

The necessary rate of lubricant supply to a hydrodynamic slider or journal bearing can be determined by the method given in Section 5.3. The inlet to the bearing must be constantly supplied with a sufficient quantity to generate the film between the surfaces and to keep the temperature to within satisfactory limits.

On the other hand, the volume of lubricant necessary to generate an elastohydrodynamic film is extremely small. In almost every case a film of fluid attaches itself to each surface and passes repeatedly through the contact, the only additional supply required being that to make up the almost negligible losses to these films. It follows, therefore, that any requirement for a large volume will result from the need to remove large quantities of heat from the vicinity of the bearing. The energy generated in the bearing is given by $P = FV$ for a linear bearing and $P = M2\pi N$ for a rotating bearing.

The heat generation in sliding bearings has been described in Chapter 5. For rolling element bearings the friction moment, $M$, may be calculated from

**Table 9.1**

| Bearing type | $f_0$ | | | $f_1$ | $g_1 P_0$ | $R$ |
|---|---|---|---|---|---|---|
| | Oil mist | Oil-bath grease | Oil bath vert. shaft | | | |
| Deep-gve. ball | 0.7–1 | 1.5–2 | 3–4 | 0.0009$Z$ | $(2.3)F_a - 0.1F_r$ | 0.85–1 |
| Self-al. ball | 0.7–1 | 1.5–2 | 3–4 | 0.0003$Z$ | $1.4YF_a - 0.1F_r$ | 0.4–0.5 |
| Ang-cont. ball single-row | 1 | 2 | 4 | 0.0013$Z$ | $F_a - 0.1F_r$ | 0.4–0.5 |
| double-row | 2 | 4 | 8 | 0.001$Z$ | $1.4F_a - 0.1F_r$ | 0.6 |
| Cyl. and needle Rs. single-row | 3–6 | 6–12 | 12–24 | 0.00025–0.0003 | $F_r \ (F_a = 0)$ | 0.5 |
| double-row | 6–10 | 12–20 | 24–40 | 0.00025–0.0003 | $F_r \ (F_a = 0)$ | 0.5 |
| spherical Rs. | 2–3 | 4–6 | 8–12 | 0.0004–0.0005 | $1.2YF_a$ | 0.3–0.4 |
| taper Rs. | 1.5–2 | 3–4 | 6–8 | 0.0004–0.0005 | $2YF_a$ | 0.3–0.4 |
| thrust balls | 0.7–1 | 1.5–2 | 3–4 | 0.0012$Z$ | $F_a$ | 0.25 |
| cyl. thrust | – | 2 | 4 | 0.0018 | $F_a$ | 0.25 |
| needle thrust | – | 2–3 | 4–6 | 0.0018 | $F_a$ | 0.25 |
| spherical thrust | – | 3–4 | 6–8 | 0.0005–0.0006 | $F_a(F_r < 0.55F_a)$ | 0.25 |

$Z = (P_0/C_0)$. Lower values are for light series, higher for heavy series.

the manufacturers' literature (e.g. SKF, 1984): $M = M_0 + M_1$, where $M_0 = f_0(60\eta N)d_m^3 \times 10^{-7}$ and $M_1 = f_1 g_1 P_0$ N·mm. $M_0$ is the moment arising from hydrodynamic losses, while $M_1$ is produced by the elastic deformation and partial localised slip between the contact surfaces. The value of $f_0$ depends on the design and lubrication, and $f_1$ and $g_1$ depend on the load and its direction. These values are given for selected cases in Table 9.1, but a fuller list is available in SKF (1984).

The energy that we require to remove from the bearing, as calculated above, will be dispersed by the usual heat transfer mechanisms of conduction, convection and radiation. The relative importance of these will depend very much on individual circumstances, but there will inevitably be some cases where we have to rely on the lubricant to carry away most of the heat. The heat thus removed can be calculated from $P = \gamma Q c T$, where $T$ is the temperature rise of the lubricant through the bearing and $c$ is the specific heat of the lubricant. For popular oils at $15\,°C$, $c$ can be taken as $1.82\,kJ/kg·°C$; for water it is about $4\,kJ/kg·°C$.

Thus, the flow rate, $Q$, can be determined. Obviously a conservative design would assume all heat removal by the lubricant.

Thermal problems are very important in bearings, especially in those rolling bearings and gears where large lubricant flows are not possible. Any increase of temperature has a detrimental effect on the viscosity, which, in turn, reduces the film thickness and may lead to an increase of heat generation. This situation may reach a stable condition or it may become disastrously unstable. This latter condition is the basis for the concept of 'flash temperature' as a criterion for failure (Blok, 1937).

Having decided on the lubricant and the necessary flow rate, we must select a supply system from the main alternatives listed below.

## 9.3   METHODS OF LUBRICANT SUPPLY

This section is confined to the supply of grease and oil to rolling bearings. These form the vast majority of lubricants used today. Which is used will depend upon the circumstances. Grease is good for keeping out contamination, operating in any attitude, and for low speeds, and is extensively used in rolling element bearings, especially in sealed units where continuous lubricant supply is not possible or is undesirable. Oil lubrication is suitable for heat removal and for hydrodynamic bearings operating at higher speeds.

### 9.3.1   Grease Supply

There are three forms of grease supply: (1) bearings packed on assembly; (2) grease nipple to each bearing for periodic replenishment; (3) grease piped to a number of bearings either by hand or by automatic pump.

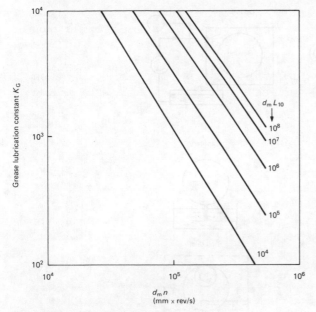

**Figure 9.1**   Grease lubrication constant

For rolling element bearings the initial charge of grease should reach the working surfaces, but the housing should not be overcharged as this can cause excessive churning and high temperatures.

The life of the bearing may be limited by the degradation of the grease unless it is replenished occasionally. The interval between regreasing is given by RHP (1977), $T = K_G R$ h, where $K_G$ is the grease lubrication constant derived from Figure 9.1 and $R$ is the relubrication factor from Table 9.1.

### 9.3.2   Oil Supply

There are many methods of supplying oil to a bearing. The most common are illustrated in Figure 9.2. Any of these methods can be found in use for rolling bearings, but for plain journal bearings and sliders, oil mists, capillaries and ring oilers are unlikely to provide sufficient flow rate.

#### Pumped Supply

In this case the oil is supplied to the bearing by a pump under pressure. The flow rate can be as large as the pump can produce and therefore the heat removal can be very effective. The capital cost, maintenance costs and initial filling cost can be high and, of course, energy is required to operate the pump. However, where a large, reliable flow is required, this system is most suitable.

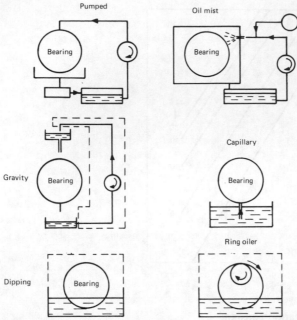

**Figure 9.2** Oil-supply systems

## Pumped Oil Mist

Mist systems (aerosol systems) produce a mist of fine lubricant droplets which travel at low velocity through the piping to a nozzle, where they are speeded up to over 40 m/s so that they wet the surfaces. Clearly, this is of use in enclosed spaces, such as gearboxes, where the mist will permeate the entire space. The air is vented to atmosphere and the oil is usually collected and recirculated. Since the thermal capacity of air is very low and the volume of oil quite small, the heat removal capability is also small. A detailed design guide to such systems is found in *The Tribology Handbook*. The initial cost and maintenance costs are relatively high.

## Gravity Supply

In cases where no great flow is necessary, gravity feeding, normally in the form of droppers, can be sufficient. This can be made automatic by collecting the oil and pumping it back to a reservoir. This system is simple and cheap, but is limited to lightly loaded, low-speed bearings.

## Capillary Systems

These systems use a wick or pad of felt in which the capillary action draws the fluid from a reservoir on to the bearing surfaces. The oil flow increases as the cross-sectional area of the wick, and decreases as the length of the wick increases and with viscosity increase. The design and construction of such systems is very simple and the costs small. However, the supply rate is quite limited and is only sufficient for lightly loaded bearings.

## Dipping Systems

If a sump is included in the assembly containing a horizontal rotating shaft, it is possible to arrange for part of the rotating mechanism to dip into the lubricant. This will carry oil up to a higher point, where it will drop onto the bearing, splash the oil onto the bearing, create an oil mist, or a combination of these actions.

It is possible for rolling element bearings to dip below the oil surface, but it is recommended that immersion be only as far as the centre of the lowest rolling element, in order to avoid excessive churning and consequent temperature rise.

The ring-oiler consists of a ring of much larger diameter than that of the shaft, which dips below the surface and as it rotates (more slowly than the shaft, obviously), carries oil by viscous lift up to the shaft. These are used for rotational speeds between 50 rev/min and 3500 rev/min and shaft sizes up to 50 mm, since the ring speed is limited to about 9 m/s. The oil supplied depends on the ring size and speed, but if more is required, it is possible to use more than one ring or to use a chain, which has a bigger surface area.

In some cases the ring can be replaced by a disc, which is convenient for low-speed bearings and high-viscosity oils.

If components dip intermittently into the oil, as in crank systems, these can be used to splash oil onto other components or to create a mist. This reduces the drag which continuously immersed components experience.

Clearly, all these methods are simple and cheap. However, it is necessary to maintain the sump level within quite close limits.

## 9.4 FILTRATION

When supplying the lubricant to the bearing, it is usually necessary to incorporate a filter in the system, to remove dirt, debris, contamination and water. The filter(s) may be located in various parts of the circuit, as shown in Figure 9.3 (which is based on *The Tribology Handbook*), according to their main purposes.

The oil reservoir vent filter is to prevent the entry of air-borne contaminants, while that on the filler is to exclude coarse solids. The pump may have a medium filter on the suction side to protect the pump and a finer one on the outlet to protect the bearings. A medium return line filter may be included to stop wear debris from entering the reservoir. In systems

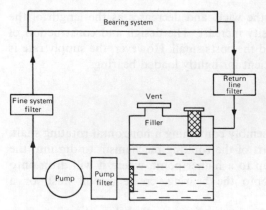

**Figure 9.3** Filtration

where cleanliness is paramount a separate bulk cleaning operation, perhaps by centrifuging the oil, may be used.

The size of the particles trapped by the filter will depend on its construction. This varies from perforated metal strainers, which will trap particles from 100 $\mu$m to 1000 $\mu$m, through woven fabrics and felts at 10–100 $\mu$m, down to ceramics and non-woven sheets at 0.5–50 $\mu$m. For specialised applications membranes are used which will remove particles down to 0.005 $\mu$m.

In addition to the normal filters, magnetic plugs can be inserted in the lines to pick up appropriate material. These consume very little energy and will collect very fine particles.

Water can be removed by using a hydrophobic element, such as silicon-coated papers or Teflon-coated meshes, and allowing the water to fall by gravity into a collection sump.

It is necessary to decide the degree of filtration necessary for any bearing. The finer the filter the higher the penalty to be paid in pressure loss through the filter or in flow rate if the pressure is preserved.

Most hydrodynamic bearings have minimum film thicknesses in the region of 20 $\mu$m upwards. Clearly, if particles of this size and above are precluded from entering the bearing, those smaller ones that remain should pass through the bearing clearance without interfering with the performance. With externally pressurised bearings the criterion will be not only the film thickness, but also the restricter diameter.

In rolling element bearings the thickness of the lubricant film is frequently much less than 1 $\mu$m. It is usually uneconomic to try to remove all the particles down to this size. The effect of degree of filtration on performance is not easy to see, but evidence suggests that filtration down to about 25 $\mu$m is beneficial; further refinement to the order of 2–3 $\mu$m produces no statistically significant improvement (Loewenthal and Moyer, 1979).

The pressure loss through a filter can be written (Wells, 1967)

$$\Delta p = \frac{Q^2 K_F \gamma}{2gA^2}$$

where $K_F$ is a characteristic of the filter, which is essentially constant in the laminar region, falling off as the flow in the filter becomes turbulent at a transition Reynolds number of 4–15, where the Reynolds number is given by

$$Re = \frac{VD}{v}$$

$D$ being usually taken as the diameter of the filter inlet port. $K_F$ is a function of the filter construction and is determined by the Kozeny–Carman relation:

$$K_F = \frac{e^3}{kS^2(1-e)^2}$$

$e$ is the void fraction or porosity, $S$ is the specific surface of solid/unit volume $(m^2/m^3)$ and $k$ is a constant for the filter material (for cellulose $k = 5.55$).

The relationship between flow rate and pressure drop is shown for a typical case in Figure 9.4, which represents the results from tests to BS 6277. This graph is for the case where the filter is equipped with a bypass which opens when the pressure drop through the filter reaches a predetermined value. This could happen if the filter were blocked.

As the pressure drop increases with $Q$, if the flow rate is high, it is often advantageous to use bypass filtration. In this case the filter is connected in parallel with the bearing system and the pressure differentials are arranged so that only a small fraction of the lubricant passes through the filter on each circuit. This means that not all the fluid is cleaned on every pass, but filtration is still quite efficient.

A more common arrangement is a combined full-flow/bypass filter, where all the lubricant is filtered to some degree, but a small proportion still goes through a bypass to remove finer contamination. Cartridges including both types of filtration in parallel in one element are available.

Since most filters operate by trapping particles within the medium, there will come a time when the trapped material will seriously interfere with the performance. With fibrous filters, where filtration operates through the depth of the element, this will be noticed as a reduction in the degree of filtration and the possible release of previously collected material due to pressure surges. For a surface filter, which operates by trapping the particles in well-defined openings on the surface, trouble is indicated by a decrease in flow rate or increase in pressure drop. In either case the filter must be serviced, by either replacing the element or cleaning it in a solvent, by steam backwash or ultrasound.

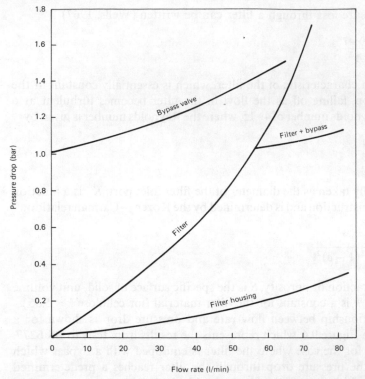

**Figure 9.4**  Pressure drop plotted against flow rate for a filter with integral bypass

## 9.5  MACHINERY HEALTH MONITORING

The traditional method of keeping machinery in a good operating condition
has been through planned preventative maintenance – i.e. regular scheduled
servicing involving oil change or replenishment, replacement of filters,
readjustment of machine settings, etc. However, this type of procedure suffers
from two drawbacks: first, to ensure good general reliability in the intervals
between services, many of the procedures are carried out more often than is
necessary; second, if a serious fault should develop between services, it could
lead to disastrous consequences before the next scheduled service. Therefore,
there is an increasing movement to overcome both these drawbacks by
replacing planned preventative maintenance by *maintenance on condition*.
Under a regime of maintenance on condition the health of the machinery is
monitored, without shutdown, either continuously or at very frequent
intervals, and maintenance procedures are effected only when necessary. The
two most effective and widely used methods of machinery health monitoring
are oil analysis and vibration monitoring. These procedures will be described
in turn.

## 9.6 OIL ANALYSIS

Techniques now exist for monitoring the condition of any oil-lubricated machine by investigating at regular intervals the amount and types of wear debris present in the oil. Samples of oil can be easily taken and the techniques do not require the machinery itself to be stripped down and examined. A sudden change in the amount or type of debris between successive oil samples may then indicate the need for immediate remedial action.

There are two types of analysis procedure in common use: spectroscopic oil analysis procedure (SOAP) and ferrography. The methods are to some extent complementary; each has its own advantages and disadvantages, which are described briefly below.

### 9.6.1 Spectroscopic Oil Analysis Procedure (SOAP)

In this procedure a small sample of oil, typically 0.1 ml, is vaporised, and its atomic constitution is determined both qualitatively and quantitatively by spectroscopic analysis. Three types of analysis are in common use: differential infrared spectroscopy; atomic emission spectroscopy; and atomic absorption spectroscopy. All give essentially the same information.

SOAP gives the total concentrations of the elements in the sample but it gives no information on whether the atoms were previously combined in a chemical compound or an alloy. Furthermore, it does not give any information on the sizes or shapes of the particles, and it detects particles only in the size range 0–3 $\mu$m. Its major advantage lies in the detection and measurement of wear debris produced by corrosive or oxidative wear, which is normally of submicron size.

### 9.6.2 Ferrography

Ferrography (Scott *et al.*, 1975) is a technique which allows metallic wear debris in an oil sample to be magnetically separated and arranged according to size on a transparent slide, so that it can be subsequently examined by optical and scanning electron microscopy and chemically analysed by electron probe microanalysis. The amount of debris and its particle size distribution can be determined by optical density measurement, and a rapid increase in the total quantity of debris or in the ratio between large and small particles is indicative of a transition to a severe wear situation.

Ferrography detects particles in the size range of 1–20 $\mu$m, and may therefore miss the very fine debris which often results from corrosive wear. The shapes of the particles are good indicators of the types of wear that are occurring. In particular, long chips of metallic debris indicate severe abrasive cutting wear, which almost certainly necessitates machine shutdown and component replacement.

The technique is under development to give on-line continuous monitoring (Centers, 1983) and to relate the particle size distribution to the state of wear (Roylance and Pocock, 1983; Roylance and Vaughan, 1983).

The information given by both SOAP and ferrography can be misleading when samples are taken from filtered systems, as the filter may permanently trap some components of the debris and thus preclude their detection and analysis. Further work is required on this and other problems before the techniques become completely reliable indicators of machine condition.

## 9.7   VIBRATION ANALYSIS

Many different types of problem can be revealed by changes in the vibration pattern of a machine, including: unbalance of rotating parts; misalignments of couplings and bearings; worn or damaged gears or bearings; looseness; and rubbing between mating surfaces.

Vibration can be measured in terms of displacement, velocity or acceleration. The two most common methods are monitoring of acceleration and displacement, for which reliable and robust accelerometers and non-contacting displacement transducers are readily available. Measurements are normally made using one or other of these systems at each bearing housing in three orthogonal directions, including the direction parallel to the axis of the bearing. However, such transducers are normally only useful up to frequencies of $\approx 2\,\mathrm{kHz}$ and cannot be used to detect higher-frequency vibrations which can occur in high-speed rolling element bearings or gearboxes. In such cases shock pulse measurements must be made.

In general, there will be slow but progressive changes in either the amplitudes or frequencies of vibration of various parts of a machine, and such changes often do not indicate serious faults. The aim of vibration monitoring is to detect either new frequencies, sudden changes in amplitude at any of the existing frequencies or significant departures from previously established trends. In any case, the frequencies at which changes are observed are often indicative not only of the onset of a fault, but also of the component in which the fault has developed. This often allows the necessary spares, tools, etc., to be made available before the machinery is shut down, thus minimising downtime.

The level of sophistication of vibration monitoring is usually dependent on the probable level of cost of any unscheduled shutdown.

At the lowest level, it usually consists of periodic, perhaps weekly or monthly, checking of the vibrations of each bearing housing with portable transducers. It is essential that such measurements are made at the same place each time, and this is usually accomplished either by marking permanent reference points on the housing or by using permanently attached blocks into which the portable transducer can be fitted reproducibly. In this type of monitoring, the results are usually recorded and plotted manually, and there is an interval of some hours, or even days, between the measurements being taken and any changes being noted and acted upon.

At the second level of sophistication, transducers are permanently mounted at critical points on machinery and the output of each transducer is monitored continuously. This output is then used in two ways. First, alarm levels are set on each output so that immediate warning will be given if any potentially dangerous vibration develops; second, the output levels are sampled at frequent intervals and trends identified by the data acquisition and processing system.

At the most sophisticated level, vibration levels are both recorded and trended continuously so that at any time the maintenance engineer has at his disposal a fully up-to-date display of the machine condition.

## 9.8  FAILURE ANALYSIS

Inevitably there will be occasions when bearings fail. It would be wrong to assume that this is always the fault of the bearing itself or the way it has been used. Problems in other machine components can produce increases of load, temperature, etc., which result in conditions for the bearing far more arduous than those for which it was designed.

When analysing bearing failure, it is essential to collect every piece of evidence available. This will be greatly helped by records from bearing monitoring devices and oil analysis, as described above, but there is no doubt that very often the most vital clues are given by examination of the failed bearing itself. The most valuable aid to this is a collection of photographs of bearing failures, with their causes. There are several such collections available to which the reader is referred for cases where failure analysis causes any difficulty (Neale, 1973; Harris, 1967; Nisbet and Mullet, 1978; SKF, 1974). In this section a brief summary will be given, to allow the more obvious cases to be identified.

### 9.8.1  Sliding Bearings

*Dirt in the Bearing*

Dirt can be introduced into the bearing either during assembly or with the lubricant during running. This will be shown by scoring and scratching of the surface in the direction of motion. Larger particles introduced during assembly can cause localised overheating and distortion.

*Inadequate Clearance in Journal Bearing, Insufficient Lubricant, Excessive Load*

Any of these will cause the film thickness to be too small for satisfactory performance. Severe damage to the bearing may result, with surface melting of the bearing material, especially if a low-melting-point alloy is used, and smearing or wiping of the surface.

*Inadequate Interference Fit, Flimsy Housing*

This will permit small oscillatory movements between the surfaces of the bearing or bearing and housing. This is a common cause of fatigue failure and is characterised by the production and oxidation of fine debris, usually reddish-brown in colour for ferrous metals, with welding or pick-up of material at the joint between the housing and bearing shell.

*Misalignment, Excessive Journal Deflection*

This is shown by uneven wear. In journal bearings it will be on diagonally opposite top and bottom halves.

*Fatigue Failure*

Fatigue damage is recognised by the pits or spalls created, where eventually pieces fall out of the surface. It can be caused by excessive dynamic loading or by cavitation.

*Corrosion*

Corrosion will differ in appearance, depending on the bearing material used. The usual sources are water, fuel combustion products, additives in the lubricant or oxidation of the lubricant.

### 9.8.2 Rolling Element Bearings

Since the predominant mechanism in rolling element failure is fatigue, the life of the bearing is defined by the number of hours that 90% of the bearings should survive (see sub-section 8.6.2). It follows, therefore, that one premature bearing failure does not indicate a bad selection or application. Neither does failure at or after the life predicted.

It is very rare for failure to result from a faulty bearing; it is almost always from poor fitting, selection or lubrication.

*Fatigue Failure*

This is the normal, expected mode of failure. It is characterised by pits of detached material. With a unidirectional load this will occur in the stationary raceway at the point directly under the load, since this is where the material suffers the highest number of high-stress applications. If this occurs repeatedly, it is likely that the load is too high. This may arise from insufficient bearing clearance to allow for thermal expansion or centrifugal effects.

*Insufficient Interference*

This causes fretting between the race and the housing, with the production of the characteristic reddish-brown debris. It is very important to get right the diametral interference of raceways, as axial clamping is an inadequate substitute.

*Misalignment, Excessive Shaft Deflection*

This will cause the normal track of dulled material on the raceways to be skewed. It does not necessarily lead to immediate failure, but can be damaging if allowed to continue. In cylindrical roller journal bearings this can lead to excessive loading on the flange, which has only very limited axial load capacity.

*Dirt in the Bearing*

Hard particles will be rolled into the bearing surface to produce dents. Grit will be crushed to powder, making the raceways become frosted in appearance. Softer dirt will produce marking of the tracks and may interfere with the free running of the elements. This will produce noise and possible wear.

*Damage During Fitting*

Since raceways are an interference fit on the shaft and in the housings, there is a great temptation to abuse them by overenthusiastic use of hammers. This can dent the races or the rolling elements and in extreme cases fracture the flanges. This damage is often visible after the bearing has run.

*Inadequate Lubrication*

This is shown by evidence of overheating, such as the parts of the bearing being blackened and showing the temper colours. The hardness of the parts will have reduced and in extreme cases the cage may melt. In grease-lubricated bearings the cage pockets and rims may be worn and the grease will be hard and dry.

## Brinelling

Dents may be present in the rolling tracks, which conform to the shape of the rolling elements. This is symptomatic of a sudden overload being applied to the bearing. This can happen during assembly, in which case the grinding marks will remain visible or when the bearing is stationary (false brinelling), which will obliterate the grinding marks.

## Corrosion

This is usually the result of the presence of water. If a small percentage of water is present in the lubricant, this can greatly reduce the fatigue life, possibly due to hydrogen embrittlement of the surfaces. Great care must be taken in the storage of bearings before use to exclude moisture.

### 9.8.3 Gears

Gear failure is not a common problem. Although damage to the surfaces may take place, running will often continue quite satisfactorily. The marking produced on the teeth by running should be examined periodically during the early life to see if a pattern is established. This can be very informative in giving clues to misalignment, excessive temperatures, etc.

## Mechanical Failure

The most dramatic failure is when a tooth breaks off. The obvious cause of this is that the tooth load produces bending stresses at the tooth root in excess of the tensile or fatigue strength of the material. For brittle materials a sudden shock load may cause the failure. The tooth may also be weakened by other forms of surface damage.

## Surface Fatigue

This is caused by the load on the tooth being too high. Pits form in the contact zone which may join together to affect large areas of the tooth.

## Scuffing

Scuffing is a result of inadequate lubrication. The lubricant film is not maintained, the temperature of the surface rises above the 'flash temperature' (Blok, 1937) and the tooth surfaces become roughened and torn. This process can lead to severe welding and complete destruction of the surfaces.

# References

Archard, J. F. (1953). *J. Appl. Phys.*, **24**, 981
Archard, J. F. (1959). *Wear*, **2**(6), 438
Archard, J. F. (1980). *Wear Control Handbook*, New York, ASME, 35
Archard, J. F. and Baglin, K. P. (1986). *Proc. I. Mech. E.*, **200**, 281
Arnell, R. D., Herod, A. P. and Teer, D. G. (1975). *Wear*, **31**, 361
Arnell, R. D., Midgley, J. W. and Teer, D. (1968). *Brit. J. Appl. Phys.*, **237**, 1543
Ashby, M. F. and Jones, D. R. H. (1986). *Engineering Materials*, vol. 2, Oxford, Pergamon Press
Barwell, F. T. (1979). *Bearing Systems – Principles and Practice*, Oxford; OUP
Barwell, F. A. (1983). *Wear*, **90**(1), 164
Bell, T., Staines, A. M., Fairhurst, W. and Semchysen, M. (1983). *Molybdenum Mosaic*, **6**(3), 14
Bitter, J. G. A. (1963). *Wear*, **6**(5), 169
Blok, H. (1937). *Second World Petroleum Congress*, Paris
Blok, H. (1952). *J. Inst. Petr.*, **38**, 673
Borik, F. (1980) *Wear Control Handbook*, New York, ASME, p. 327
Bowden, F. P. and Tabor, D. *The Friction and Lubrication of Metals*, Pt. 1 (1950) and Pt. 2 (1964); Oxford, OUP
Bragg, W. L. (1924). *Introduction to Crystal Analysis*, London, Bell
Briggs, G. A. D. and Briscoe, B. J. (1975). *Wear*, **35**, 337–364
Buckley, D., *see* Miyoshi and Buckley (1985), p. 18
Burwell, J. T. (1957). *Wear*, **1**, 119
Cameron, A. (1981). *Basic Lubrication Theory*, Chichester, John Wiley
Cameron, A. and Wood, W. L. (1949). *Proc. I. Mech. E.*, **161**, 59–72
Carslaw, H. S. and Jaeger, J. C. (1959). *Conduction of Heat in Solids*, Oxford, Clarendon
Centers, P. W. (1983). *Wear*, **90**, 1–9
Challen, J. M. and Oxley, P. L. B. (1979). *Wear*, **53**, 229
Challen, J. M. and Oxley, P. L. B. (1984). *Int. J. Mech. Eng. Science*, **26**, 403
Childs, T. H. C. (1988). *Proc. I. Mech. E.*, **202**(C6), 379
Cole, J. A. and Hughes, C. J. (1956). *Proc. I. Mech. E.*, **170**, 499–510
Dearnaley, G. and Grant, W. A. (1985). In *Recent Developments in Surface Coatings and Modification Processes*, London, Mechanical Engineering Publications, p. 43
Dowson, D. and Higginson, G. R. (1959). *J. Mech. Eng. Science*, **1**(1), 6
Dowson, D. and Miranda, A. A. S. (1975). *Conference on Cavitation and Related Phenomena*, University of Leeds/Mechanical Engineering Publications

Dowson, D. and Taylor, C. M. (1975). *Conference on Cavitation and Related Phenomena*, University of Leeds/Mechanical Engineering Publications

Dowson, D. and Whomes, T. L. (1967). *Proc. I. Mech. E.*, **182**, Pt. 1, No. 14

Dubois, G. B. and Ocvirk, F. W. (1953). *NACA Report No. 1157*, Washington DC

Dyson, A. (1975). *Trib. Int.*, June, 117

Earles, S. W. and Powell, D. G. (1971). *Wear*, **18**, 381

Edwards, C. M. and Halling, J. (1968). *J. Mech. Eng. Science*, **10**, 101

El-Shafei, T. E. S., Arnell, R. D. and Halling, J. (1983). *ASLE Trans.*, **26**(4), 481–487

Endo, F. and Fukuda, Y. (1965). *Proc. 8th Japan Congress on Testing and Materials*, Tokyo, p. 51

Evans, C. R. and Johnson, K. L. (1986). *Proc. I. Mech. E.*, **200**, 313

ESDU (1965). *General Guide to the Choice of Journal Bearing Type*, Item No. 65007, London

ESDU (1967). *General Guide to the Choice of Thrust Bearing Type*, Item No. 67033, London

ESDU (1977). *Low Viscosity Process Fluid Lubrication of Journal Bearings*, Item No. 77013, London, revised 1982

ESDU (1984). *Calculation Methods for Steadily Loaded Axial Groove Hydrodynamic Journal Bearings*, Item No. 84031, London

ESDU (1985). *Film Thickness in Lubricated Hertzian Contacts (EHL): Part I*, Item No. 85027, London

ESDU (1987). *Guide to the Design and Material Selection for Dry Rubbing Bearings*, Item No. 87007, London

ESDU (1988). *Selection of Alloys for Hydrodynamic Bearings*, Item No. 88018, London

Ferrante, J., Smith, J. R. and Rose, J. H., *see* Miyoshi and Buckley (1985), p. 138

Finnie, I., Levy, A. and McFadden, D. H. (1979). In Adler, W. F. (ed.), *Erosion: Prevention and Useful Applications*, Philadelphia, ASTM-STP-664, p. 36

Floberg, L. (1961). *Trans. Chalmers Univ. of Tech.*, 234

Glaeser, *see* Miyoshi and Buckley (1985), p. 214

Godfrey, D. (1980). *Wear Control Handbook*, New York, ASME

Gohar, R. (1988). *Elastohydrodynamics*, Chichester, Ellis Horwood

Gohar, R. and Cameron, A. (1966). *Proc. ASME/ASLE Conf.*, Minneapolis

Gough, V. E. (1954). *Auto Eng.*, April

Green, A. P. (1955). *Proc. Roy. Soc.*, **228A**, 191

Greenwood, J. A. (1984). *Proc. Roy. Soc.*, **393A**, 133

Greenwood, J. A. and Halling, J. (1971). *I. J. Mech. E. E.*, **1**(2), 35

Greenwood, J. A. and Williamson, J. B. P. (1966). *Proc. Roy. Soc.*, **295A**, 300

Grubin, A. N. (1949). *Central Scientific Research Institute of Technology and Mechanical Engineering*, Book No. 30, Moscow, DSIR Translation No. 337

Halling, J. (1975a). *Principles of Tribology*, London, Macmillan Education, p. 308

Halling, J. (1975b). *Wear*, **34**, 239

Halling, J. (1986). *Proc. I. Mech. E.*, **200**(C1), 31

Halling, J., Arnell, R. D. and Nuri, K. A. (1988). *Proc. I. Mech. E.*, **202**(C4), 269

Halling, J. and Nuri, K. A. (1974). *IUTAM Symposium*, University of Delft Press

Halling, J. and Nuri, K. A. (1985). *Proc. I. Mech. E.*, **199**(C2), 139

Hamilton, G. M. (1983). *Proc. I. Mech. E.*, **197**, 53

Hammitt, F. G. (1980). *Wear Control Handbook*, New York, ASME

Hamrock, B. J. and Brewe, D. E. (1983). *J. Lub. Tech.*, **105**, 171

Hamrock, B. J. and Dowson, D. (1981). *Ball Bearing Lubrication*, John Wiley

Harris, J. H. (1967). *Lubrication of Rolling Bearings*, London, Shell-Mex and BP

Harris, T. A. (1984). *Rolling Bearing Analysis*, Chichester, John Wiley

den Hartog, J. P. (1952). *Strength of Materials*, New York, McGraw-Hill

Hays, D. F. (1959). *J. Basic Eng.*, March, 13–23

Heathcote, H. L. (1921). *Mech. World*, **70**(1804), 79

Herrebrugh, K. (1968). *Trans. ASME Series F*, 262

Hitchman, M. L. (1985). In *Recent Developments in Surface Coatings and Modification Processes*, London, Mechanical Engineering Publications

Hokkirigawa, K. and Li, Z. Z. (1987). *Proc. Conf. Wear Mat.*, Houston, ASME, p. 585

Holmberg, K. (1982). *Trib. Int.*, June, 123

Holmes, R. (1963). *Proc. I. Mech. E.*, **177**, 291–307

Honeycombe, R. W. K. (1968). *The Plastic Deformation of Metals*, Arnold

Honeycombe, R. W. K. (1980). *Steels: Microstructure and Properties*, Arnold

Hutchings, I. (1979). *Erosion: Prevention and Useful Applications*, ASTM-STP-664, p. 59

Jaeger, J. C. (1942). *Proc. Roy. Soc. NSW*, **56**(203), 378

Jahanmir, S., Suh, N. P. and Abrahamson, E. R. (1974). *Wear*, **28**, 235

Jakobsson, B. and Floberg, L. (1957). *Trans. Chalmers Univ. of Tech.*, No. 3

Jakobsson, B. and Floberg, L. (1958). *Trans. Chalmers Univ. of Tech.*, No. 5

Johnson, K. L. (1968). *J. Mech. Phys. Solids*, **16**, 395

Johnson, K. L. (1970). *J. Mech. Eng. Science*, **12**(1), 9

Johnson, K. L. (1985). *Contact Mechanics*, Oxford, Clarendon

Kannel, J. W. (1965). *Proc. I. Mech. E.*, **180**(3B), 135

Kragelsky, I. V. (1965). *Friction and Wear*, London, Butterworths

Kruschov, M. M. (1957). *Proc. Conf. Lub. Wear*, London, I. Mech. E., p. 655

Lanes, R. F., Flack, R. D. and Lewis, D. W. (1982). *Trans. ASLE*, **25**(3), 289–298

Lim, S. C. and Ashby, M. F. (1987). *Acta Metall.*, **35**, 1

Loewenthal, S. W. and Moyer, D. W. (1979). *J. Lub. Tech.*, **101**, April, 171

Malik, M., Chandra, M. and Sinhasan, R. (1982). *Trans. ASLE*, **25**(1), 133–140

Marsh, H. (1964). *The Stability of Aerodynamic Gas Bearings*, Mechanical Science Monograph No. 2, London, Institute of Mechanical Engineers, pp. 1–44

Martin, F. A. (1970). *Proc. Inst. Mech. Eng.*, **184**, Pt. 3L, 120–138

Martin, F. A. (1983a). *Trib. Int.*, **16**(2), 64–65

Martin, F. A. (1983b). *Trib. Int.*, **16**(3), 147–164

Martin, H. M. (1916). *Engineering*, **102**, 199

Massey, B. S. (1989). *Mechanisms of Fluids*, London, Van Nostrand Reinhold, Chapt. 6

Miyoshi, K. and Buckley, D. H. (1985). In Loomis, W. R. (ed.), *New Directions in Lubrication, Materials, Wear and Surface Interactions*, New Jersey, Noyes Publications, p. 282

Moes, H. and Bosma, R. (1971). *Trans. ASME, J. Lub. Tech.*, **93**, 302–306

Morris, J. A. (1967). In *Lubrication and Lubricants*, Lausanne, Elsevier, p. 310

Neale, M. J. (1967). *Proc. I. Mech. E.*, **182**(3A), 547

Neale, M. J. (ed.) (1973). *The Tribology Handbook*, London, Newnes-Butterworths

Nisbet, T. S. and Mullet, G. W. (1978). *Rolling Bearings in Service*, London, Hutchinson Benham

OECD (1968). *Glossary of Terms and Definitions in the Field of Friction, Lubrication and Wear*, Paris

Peterson, H. B. and Winer, W. O. (eds) (1980). *The Wear Control Handbook*, New York, ASME

Petrusevich, A. I. (1951). *Izd. Akad. Nauk SSSR*, OTN 2, 209

Purday, H. F. P. (1949). *Streamline Flow*, London, Constable

Quinn, T. J. F. (1971). *Wear*, **18**, 413

Rabinowicz, E. (1970). In *Friction and Lubrication in Deformation Processing*, New York, ASME, pp. 90–102

Ramalingan, S. (1980). In *Wear Control Handbook*, New York, ASME

RHP (1977). *Standard Bearings Technical Catalogue*, Chelmsford

Roelands, C. J. A. (1966). *Druk. VRB Groningen*

Rowe, C. N. (1980). In *The Wear Control Handbook*, New York, ASME, p. 143

Roylance, B. J. and Pocock, G. (1983). *Wear*, **90**, 113

Roylance, B. J. and Vaughan, D. A. (1983). *Wear*, **90**, 137

Sasada, T., Norose, S. and Mishina, H. (1979). *Proc. Int. Conf. Wear Materials*, Dearborn, MI, ASME, p. 72

Savage, R. H. (1948). *J. Applied Physics*, **19**, 1

Schmitt, G. F. (1980). In *The Wear Control Handbook*, New York, ASME

Scott, D., Seifert, W. W. and Westcott, V. C. (1975). *Wear*, **34**, 251

Sherbiney, M. A. and Halling, J. (1976). *Wear*, **40**, 325

SKF (1974). *Rolling Bearing Damage: A Morphological Atlas*, King of Prussia, Pa., The Centre

SKF (1984). *Principles of Bearing Selection and Application*, King of Prussia, Pa., The Centre

Soda, N., Kimura, Y. and Tanaka, A. (1977). *Wear*, **43**(2), 165

Stribeck, R. (1901), *Z. ver. D. I.*, **45**(3), 73–125

Suh, N. P. (1973). *Wear*, **25**, 111

Suh, N. P. (1977). *Wear*, **44**, 16

Tabor, D. (1985). In Loomis, W. R. (ed.), *New Directions in Lubrication, Materials, Wear and Surface Interactions*, New Jersey, Noyes Publications, p. 2

Tallian, T. E. (1967). *Trans. ASLE*, **10**, 418

Tavernelli, J. F. and Coffin, L. F. (1959). *Trans. Am. Soc. Metals*, **51**, 438

Teer, D. G. and Arnell, R. D. (1985). In *Recent Developments in Surface Coating and Modification Processes*, London, Mechanical Engineering Publications, p. 21

Walls, J. M. (ed.) (1989). *Methods of Surface Science*, Cambridge, CUP

Waterhouse, R. B. (1988). *Fretting Corrosion*, Oxford, Pergamon Press

Weber, C. and Sallfeld, K. (1954). *Z. angew. Math. Mech.*, **34**, Nos. 1–2, 54

Wells, R. M. (1967). *Proc. I. Mech. E.*, **182**, Pt. 3A, 443

Wilson, R. W. (1975). Proc. Eurotrib. 1, *J. Mech. Eng.*, 165

Woodruff, D. P. and Delchar, T. A. (1988). *Modern Techniques of Surface Science*, Cambridge, CUP

Woolacot, R. G. (1965). *National Engineering Laboratory Report No. 194*, East Kilbride, NEL

Woolacot, R. G. and Macrae, D. (1967). *National Engineering Laboratory Report No. 315*, East Kilbride, NEL

Wymer, D. G. and Cameron, A. (1974). *Proc. I. Mech. E.*, **188**, 221

Zum Gahr, K. H. (1987). In *Microstructure and Wear of Materials*, Amsterdam, Elsevier, Chapt. 5

# Index